Tristan Visentin

Polarimetric Radar for Automotive Applications

Karlsruher Forschungsberichte
aus dem Institut für Hochfrequenztechnik und Elektronik

Herausgeber: Prof. Dr.-Ing. Thomas Zwick

Band 90

Eine Übersicht aller bisher in dieser Schriftenreihe erschienenen Bände finden Sie am Ende des Buches.

Polarimetric Radar for Automotive Applications

by
Tristan Visentin

KIT Scientific Publishing

Karlsruher Institut für Technologie
Institut für Hochfrequenztechnik und Elektronik

Polarimetric Radar for Automotive Applications

Zur Erlangung des akademischen Grades eines Doktor-Ingenieurs von der KIT-Fakultät für Elektrotechnik und Informationstechnik des Karlsruher Instituts für Technologie (KIT) genehmigte Dissertation

von M.Sc. Tristan Visentin

Tag der mündlichen Prüfung: 7. Dezember 2018
Referent: Prof. Dr. Thomas Zwick
Korreferent: Prof. Dr. Christian Waldschmidt

Impressum

KIT Scientific Publishing

Karlsruher Institut für Technologie (KIT)
KIT Scientific Publishing
Straße am Forum 2
D-76131 Karlsruhe

KIT Scientific Publishing is a registered trademark of Karlsruhe Institute of Technology.
Reprint using the book cover is not allowed.

www.ksp.kit.edu

Print on Demand 2019 – Gedruckt auf FSC-zertifiziertem Papier

ISSN 1868-4696
ISBN 978-3-7315-0888-5
DOI 10.5445/KSP/1000090003

Preface of the Editor

In recent years, radar sensors have established a stronghold in the automotive industry and therefore especially in Germany a large number of automotive radar sensors will be developed and manufactured. A variety of studies have shown that by the year 2030, the percentage of vehicles in use worldwide, fitted with at least one radar sensor, will range between 30 % and 80 %. At present, the automotive radars predominantly serve as a part of advanced driver assistance systems, i.e., the responsibility lies exclusively with the driver. This scenario is bound to change in the near future with the introduction of self-driven cars on our roads. In a fully autonomous vehicle, the responsibility of every single driving maneuver will lie with the system, which poses an enormous challenge to the reliability of the sensors involved. Due to the extreme complexity of the vehicular environment, a redundancy of sensors is inevitable. Consequently, all car manufacturing companies agree on the fact that a multitude of sensor technologies (radar, camera, and perhaps lidar and ultrasonic) should be simultaneously deployed in vehicles and several sensors of the same technology should be used to scan a given environment. In addition, the performance of sensors needs to be improved as well. In the field of radars, the recent technological advances focus towards transmit and receive channel systems, and multi-static systems. In this regard, the underlying potential of radar polarization has not been exploited to date. This is exactly where the work of Dr. Tristan Visentin comes into play. In his dissertation, he has worked upon the scientific principles involved in the application of polarimetry in automotive radars. He has built a dual-polarized radar system and investigated its advantages and disadvantages over a conventional system, in both theory and practice. Mr. Visentin has developed a fully polarimetric antenna system for an 8×8 multiple-input, multiple-output (MIMO) radar. In addition, he has selected a suitable calibration process for the radar and verified it as well. The field tests carried out using the new measurement system have led to novel findings regarding the untapped potential of polarimetry in automotive radars. For the first time, Mr. Visentin has used Deep Learning Algorithms for the classification of objects in a given road traffic scenario, based on polarimetric measurements, thereby demonstrating the fundamental feasibility of the concept. In his work, Dr. Tristan Visentin presents important results about the use of polarimetry in automotive radar sensors. I am sure that his work will draw much attention worldwide. For Dr. Visentin, with his broad technical skills, I wish him further much success in his career.

Prof. Dr.-Ing. Thomas Zwick

- Institute Director -

Polarimetric Radar for Automotive Applications

Zur Erlangung des akademischen Grades eines

DOKTORS DER INGENIEURSWISSENSCHAFTEN
(Dr.-Ing.)

von der KIT-Fakultät für

Elektrotechnik und Informationstechnik des
Karlsruher Instituts für Technologie (KIT)

angenommene

DISSERTATION

von

M.Sc. Tristan Visentin

Tag der mündlichen Prüfung:	7. Dezember 2018
Referent:	Prof. Dr. Thomas Zwick
Korreferent:	Prof. Dr. Christian Waldschmidt

Kurzfassung

Die vorliegende Arbeit stellt die erste tiefgehende Untersuchung und die grundlegenden Prinzipien der Radarpolarimetrie für automotive Anwendungen dar. Als Basis für die messtechnische Untersuchung wurde ein experimenteller frequency-modulated continuous-wave (FMCW) Radarsensor mit einer orthogonal linear polarisierten Hohlleiterantenne, bestehend aus Sektorhornelementen, ausgestattet. Unter Verwendung dieser Antenne in Kombination mit einem time-division multiplexing (TDM) multiple-input, multiple-output (MIMO) Ansatz wurde ein vollpolarimetrischer und quasi echtzeitfähiger FMCW Radarsensor bei 77 GHz aufgebaut. Ein neuartiger Ansatz von überlappenden Empfangs- und Sendekanalkombinationen mit orthogonalen Polarisationen ermöglichte hierbei die kohärente Messung der gesamten komplexen Streumatrix, unabhängig von der Zielposition.

Nachfolgend wurde ein bestehendes Verfahren zur Antennenkalibration verallgemeinert, so dass auch bistatische und quasi-monostatische MIMO Radare kalibriert werden können, wenn diese Radare Antennen hoher Polarisationsreinheit aufweisen. Zur Verifikation der Kalibrierung wurden Messungen von kanonischen Objekten (z.B. Doppel- und Tripelspiegel) aus verschiedenen Richtungen in einer Absorbermesskammer durchgeführt, um sowohl die Kalibrationsmethode zu verifizieren als auch die auftretenden polarimetrischen Effekte bei Millimeterwellen aufzuzeigen. Auf dieser Basis wurden kohärente und inkohärente Zerlegungsmethoden der Streumatrix, welche bisher nur in der synthetic aperture radar (SAR) Polarimetrie verwendet wurden, erfolgreich auf die Messdaten angewendet.

In einem weiteren Schritt wurde eine gründliche Untersuchung von Mehrwege- und Richtungsdetektion durchgeführt. Mehrwegereflexe, die durch metallische Oberflächen entstehen, ändern den Polarisationszustand und wurden daher mit einem Messaufbau in der Messkammer untersucht. Anhand der Ergebnisse wurden vielversprechende Ansätze ausgearbeitet, um die Richtungsdetektion in Gegenwart von Mehrwegeempfang zu verbessern.

In einem weiteren Schritt wurde das Zielklassifizierungspotential von polarimetrischen Daten untersucht. Zu diesem Zweck wurden mehrere Zweiräder

als Repräsentationen von verwundbaren Verkehrsteilnehmern in der Messkammer auf einem Drehtisch vermessen. Mit dieser Methode konnten polarimetrische Rotationssignaturen der Fahrzeuge erstellt werden. Mittels der kohärenten Pauli-Zerlegung wurden farbkodierte Distanz-Azimut Radarbilder erstellt. Diese zeigten individuelle räumliche Formen und Farbverläufe der einzelnen Fahrzeuge über dem Aspektwinkel und bewiesen damit das Potential der hohen Klassifizierungsmöglichkeit.
Diesem Gedanken folgend wurde maschinelles Lernen in Form von convolutional neural networks (CNNs) verwendet, um die Klassifizierung durchzuführen. Dafür wurden mehrere künstliche Ziele, die eine variierende Anzahl an kanonischen Objekten aufwiesen, hergestellt. Ähnlich wie die Zweiräder wurden diese Ziele auf einem Drehtisch vermessen. Es stellte sich heraus, dass Trainingsdaten der Ziele in verschiedenen Distanzen vorliegen müssen, damit das optimierte CNN eine hohe Generalisierung erreicht. Mit diesem Ansatz konnte eine Klassifizierungsgenauigkeit von über 90 % erreicht werden.
Um reale Feiluftmessungen durchführen zu können wurde ein fahrzeugintegrierbarer Sensor aufgebaut. Dazu wurde eine zweite dual-orthogonal polarisierte Hohlleiterantenne mit doppelter Aperturgröße in Azimut aber gleicher Anzahl an Antennenelementen entworfen. Diese Antenne wurde mittels eines innovativen Fertigungsprozesses aus kupferbeschichteten stereolitographisch 3D-gedruckten Schichten angefertigt.
Schließlich wurde ein experimentelles FMCW Radar mit der verbesserten Antenne ausgestattet und für reale Freiluftmessungen von statischen und verkehrsrelevanten Zielen (u.a. Auto, stehendes und liegendes Motorrad und Fahrrad) auf einer Teststrecke verwendet. Mittels gemessener Radarbilder und einer erzeugten ground truth (GT) der Messszene wurden CNNs zur Zielklassifizierung angewandt. Es konnte gezeigt werden, dass ein vollpolarimetrisches Radar eine um ca. 10 % gesteigerte Klassifizierungsgenauigkeit gegenüber einem einfach polarisierten Radar mit der gleichen räumlichen Auflösung aufweist. Nach Filterung von verrauschten Daten konnte eine max. Klassifizierungsgenauigkeit von ca. 90 % mit dem vollpolarimetrischen Ansatz erreicht werden. Noch bedeutender scheint, dass ein vollpolarimetrisches MIMO Radar mit der gleichen Anzahl an realen Antennenkanälen, aber einer um fast dreimal kleineren Apertur im Vergleich zu dessen einfach-polarisierten Gegenstück, eine vergleichbare Klassifizierungsgenauigkeit aufweist.

Abstract

This thesis presents the first in-depth analysis and fundamental approaches of radar polarimetry for automotive applications. As basis for the experimental investigation, an experimental frequency-modulated continuous-wave (FMCW) radar sensor was equipped with an orthogonal linearly-polarized waveguide antenna comprising sectoral horn elements. Using this antenna in combination with a time-division multiplexing (TDM) multiple-input, multiple-output (MIMO) approach, a fully polarimetric and quasi real-time capable FMCW radar sensor operating at 77 GHz was built. A novel approach of overlapping receiver-transmitter channel combinations with orthogonal polarizations enabled the coherent measurement of the full complex scattering matrix (S-matrix), independent of the target position.
Next, an existing calibration method was generalized in order to allow also calibration of bistatic and quasi-monostatic MIMO radars operating with antennas of high polarization purity. Multi-angular measurements in an anechoic chamber were performed on canonical objects (like dihedral and trihedral corner reflectors) to prove the calibration method and to demonstrate the occurring polarimetric effects in millimeter wave (MMW) range. On this basis, coherent and incoherent decompositions of the S-matrix were successfully applied to the measurement data, which had only been used in synthetic aperture radar (SAR) polarimetry before.
In a next step, a rigorous analysis of multipath and direction of arrival (DOA) detection, focusing on MMW radar polarimetry, was conducted. Multipaths originating from metallic surface reflections change the polarization state and were thus investigated with a measurement setup in an anechoic chamber. From the measurement results, promising approaches to improve DOA detection in the presence of multipath reception were elaborated.
In a further step, the target classification potential using polarimetric data was investigated. For this purpose, several two-wheeled vehicles, as representations of vulnerable road users, were measured in an anechoic chamber on a rotary table. In this way, polarimetric rotation signatures of the vehicles could be generated. By using the coherent Pauli decomposition method, color-coded

radar images were created. These results illustrated the unique spatial shapes and color gradients of the different vehicles over aspect angle and confirmed the potential of high target classification capability.
Following this idea, a machine learning approach using convolutional neural networks (CNNs) to perform the classification task was applied. For this purpose, several artificial targets that contained varying numbers of canonical objects, were manufactured. Rotation measurements on a rotary table were performed with these targets, in the same way as with the two-wheeled vehicles. It was found out that training data for different target distances is required in order to reach a good generalization of the optimized CNN. With this novel approach, a high classification accuracy of more than 90 % could be achieved.
In order to perform real world outdoor measurements, a vehicle-integrable radar sensor was built. For this sensor, a second dual-orthogonally polarized waveguide antenna was designed with double virtual azimuthal aperture size, but the same number of sectoral horn elements as the first one. This antenna was manufactured using an innovative stereolithography apparatus (SLA) 3D printed and copper-coated layer stack technique.
Finally, an experimental FMCW radar was equipped with the improved antenna, and was used for real world outdoor measurements of stationary and traffic relevant targets (including among others a car, standing and lying motorbike and bicycle) on a test track. Using the resulting radar images and a created ground truth (GT) of the scenery, CNNs were applied to classify the targets. It could be shown that a fully polarimetric radar is advantageous in terms of target classification accuracy by approx. 10 %, compared to a single-polarized one with the same spatial resolution. After filtering of noisy data, a top classification accuracy of approx. 90 % could be reached with the fully polarimetric approach. Even more important seems that a fully polarimetric MIMO radar with the same number of hardware antenna channels, but almost three times smaller azimuthal aperture size compared to its single-polarized counterpart, performed equivalently in terms of target classification accuracy.

Danksagung

Die Betreuung meiner Dissertation wurde vom Institut für Hochfrequenztechnik und Elektronik (IHE) bereitgestellt. Insbesondere möchte ich dort Herrn Prof. Thomas Zwick für die Betreuung der Dissertation und die Unterstützung bei Veröffentlichungen danken. Auch danke ich Prof. Werner Wiesbeck und allen Mitarbeitern am IHE, die mit anregenden Diskussionen zum Erfolg der Dissertation beigetragen haben. Weiterhin möchte ich einen besonderen Dank an Prof. Christian Waldschmidt vom Institut für Mikrowellentechnik der Universität Ulm für die Begutachtung meiner Dissertation und die fachlichen Diskussionen am Anfang meiner Dissertationsphase aussprechen.
Da ich die Ergebnisse meiner Dissertation bei der Firma Bosch am Forschungscampus in Renningen erzielte, widmet sich dieser Teil der Danksagung an die Personen bei Bosch. Ein ganz besonderer Dank gilt meinem Betreuer Dr. Jürgen Hasch für die fachliche Betreuung der Dissertation und die Unterstützung bei Veröffentlichungen und allen anderen Dingen im Firmenalltag. Ein spezieller Dank gilt auch meinem Gruppenleiter Dr. Thomas Binzer für seine Unterstützung und Förderung in allen Bereichen, u.a. bei Einstellung meiner Studenten, der Finanzierung meiner Bestellungen und meiner Konferenzbesuche, fanden wir immer gemeinsam eine Lösung, selbst bei Schwierigkeiten. Ein weiterer spezieller Dank gilt Dr. Xuyang Li und Dr. Markus Gonser, die mir beide finanziell und fachlich mit Ihren Forschungsprojekten zur Seite standen und mir somit ermöglicht haben meine Resultate zu erzielen. Sowohl an Dr. Werner Sörgel, als auch an alle anderen Mitarbeiter der Forschungsgruppe für Hochfrequenztechnik und EMV (ARE1), die mir u.a. mit anregenden Diskussionen und bei meinen Laboraufbauten stets zur Seite standen, geht ein besonderer Dank. Des Weiteren möchte ich mich bei Rossen Michev, Andrej Sagainov, Daniel Rusev und Bilal Sultan, die ich im Rahmen der Dissertation bei Ihren studentischen Abschlussarbeiten betreut habe, für Ihre erfolgreichen Arbeiten bedanken. Ihr habt mit euren Ergebnissen einen wesentlichen Beitrag zum Erfolg dieser Dissertation geleistet. Weiterhin möchte ich auch den Gruppen für Konstruktion und der Werkstatt (ARG1 und ARG3) einen ganz besonderen Dank aussprechen. Ich danke hi-

ermit Dirk Schmidt, Dr. Ralf Mendgen, Dieter Amesoeder, Igor Savintsev, Alf Hermann, Uwe Prang und allen, die ihrerseits an der zügigen und qualitativen Fertigstellung meiner Anfragen und Auträge beteiligt waren. Zuletzt möchte ich noch Dr. Michael Pfeiffer, Dr. Kilian Rambach und Kanil Patel aus der Forschungsgruppe für künstliche Intelligenz (PJ-AI-R) für Ihre Unterstützung im Bereich Deep Learning und die zahlreichen fachlichen Diskussionen danken.
An meine Familie und meine Freunde möchte ich natürlich auch einen ganz speziellen Dank richten. Ohne euch hätte es diese Dissertation nie gegeben. Insbesondere danke ich meiner Familie, die mich stets bei meinen Zielen und im Privaten unterstützt und an mich geglaubt hat. Ich danke auch all meinen Freunden und meiner Freundin, die mich mental und wo es nur ging unterstützt und motiviert haben.

Stuttgart, 07. Mai 2018 *Tristan Visentin*

Contents

1 Introduction **1**
1.1 Motivation 1
1.2 State of the Art and Literature Review 2
1.3 Outline of Thesis 4

2 Basics of Radar Polarimetry and MIMO Radar **7**
2.1 Plane Waves and Polarization 7
2.1.1 Jones Vector 8
2.2 Polarimetric Scattering Matrix 9
2.3 MIMO Radar 11
2.3.1 Calculation of Virtual Array Positions 13
2.3.2 DOA estimation for an 1D-ULA 14
2.4 Speckle, Glint & Scintillation 15
2.5 Canonical Objects 17
2.5.1 Canonical Objects in Real Environments 21
2.6 Polarimetric Decompositions 23
2.6.1 Coherent Decompositions 24
2.6.2 Incoherent Decompositions 26

3 Experimental Fully Polarimetric MIMO FMCW Radar System at 77 GHz **35**
3.1 Block Diagram 35
3.2 8x8 Dual-Polarized Waveguide Antenna 36
3.2.1 Virtual MIMO Array 37
3.2.2 Full-Wave Simulation and Antenna Pattern Measurements 39
3.3 Redesign of 8x8 Dual-Polarized Waveguide Antenna 42

4 Polarimetric Calibration Methods **47**
4.1 Calibration Based on the IACT 47
4.2 General Calibration Techniques 54

5 Fixed Sensor Multi-Angle Meas. in Anech. Chamber 57
5.1 Statistical Evaluation of Arbitrarily Located Canonical Objects 57
5.2 Rotational Measurements of Canonical Objects 59
5.3 Measurements of Targets on Rotary Table 66
5.3.1 Rotated Two-Wheeled Vehicles 66
5.4 Analysis of Multipath and DOA Detection 72
5.4.1 Multipaths Over Roads and Guardrails 72
5.4.2 Measurement Results 76

6 Classification of Polarimetric Measurement Data Based on Deep Learning 89
6.1 Convolutional Neural Networks (CNNs) 89
6.2 CNN Architecture and Training Process 90
6.3 Polarimetric Targets for Classification 91
6.4 Measurements and Data Preprocessing 93
6.5 Training and Classification Results 93
6.6 Approaches for Classification Performance Improvement . . . 99
6.6.1 Normalization Techniques 100
6.6.2 Comparison of Normalization Methods and Usage of an Additional Training Set 101

7 Classification of Stationary Targets in Vehicular-Based Measurements 105
7.1 Signal Processing Flow Chart 106
7.2 Test Track Scenario and Measurement Vehicle Setup 107
7.3 Data Preprocessing 109
7.3.1 Fully Polarimetric and Single-Polarized VAs with Equal Channel Count 109
7.4 Ground Truth Generation 112
7.5 ROI Extraction 114
7.6 Dataset Analysis 116
7.6.1 Low Dimensional Data Projections 118
7.6.2 Evaluation of Fully Polarimetric Data 119
7.6.3 Comp. of Fully Polarimetric and Single-Polarized Data 125
7.7 Classification of Targets Using Deep Learning 128
7.7.1 Comparison of Global and Local Dataset Normalization 130
7.7.2 Classification Results Using a Filtered Dataset 132

8 Summary and Conclusion **135**

A Appendix **141**
A.1 Fast Chirp FMCW Radar Principle and Digital Signal Processing 141
A.2 FMCW Modulation Parameters 144

Abbreviations and Acronyms **147**

1 Introduction

Radar which is short for "**ra**dio **d**etection **a**nd **r**anging" and lidar (short for "**li**ght **d**etection **a**nd **r**anging") are two of the most demanded key sensing technologies complementing camera systems for the myriad tasks of autonomous driving [RV17]. Whereas current automotive radar sensors prove to be a weather robust, low-cost and solid state solution, state-of-the-art automotive lidars are costly, suffer from various mechanical moving parts and are susceptible to adverse weather conditions (e.g. heavy rain, snow, and fog), due to their optical frequencies [LaP17b]. On the other hand, lidars are currently the only high-performance sensor systems that provide very high resolution in 3D and thus can precisely map the surrounding [Che17]. Currently, automotive radar sensors are still suffering from low resolution and are thus only capable of very basic classification of the detected targets [LaP17a]. Imaging and cognitive radars that can not only detect and range but perceive and understand their environments are therefore becoming ever increasingly important [Hay06; Dic+16].
The huge market boost in the autonomous driving sector is the main driver of the advanced driver assistance systems (ADAS) market. The global autonomous driving market is expected to be worth 83 billion USD by the year 2025 [Sha17], from which the global ADAS market will have an estimated share of 67.43 billion USD growing at a compound annual growth rate (CAGR) of 19 % [Tre18]. From all the ADAS, the global market share of automotive radar modules is expected to reach 12.16 billion USD by the year 2025, growing at a high CAGR of 20.8 % [Gra17]. This constitutes approx. 18 % of the total ADAS.

1.1 Motivation

These numbers give an idea of the significant role of automotive radars. It is apparent that the research field of imaging and high resolution 77 GHz radars that can sense, perceive and understand (classify) their environment has thus become one of the most outstanding. In that regard, it is surprising that not much research has yet been conducted on polarimetric radars for automotive applications. This type of radar uses orthogonal polarizations on transmit and receive. Hereby, it provides up to five times more information about the scenery as currently used single polarized radars. Polarimetric radars are therefore being deployed in a growing number of fields. The additional polarimetric information has been used beneficially in synthetic aperture radar (SAR) earth observation, radar imaging, ground penetrating radar (GPR) subsurface scanning, Doppler weather radar, and road surface monitoring. In SAR polarimetry, the extra scattering information is utilized for terrain classification and soil moisture estimation [LP09]. In polarimetric radar imaging, many features of the observed targets appear that remain concealed in single-polarized data [AS13]. Also, polarimetry has been successfully applied for classification of buried GPR targets, road surface conditions, and road debris [Yu+17; KD94; SL00]. Furthermore, polarimetric (Doppler) weather radars provide additional information which allow to identify precipitation types [BC01]. These examples illustrate the potential and importance of polarimetry for radar sensing. On the basis of the successful state-of-the-art applications, this thesis presents the first thorough analysis and groundbreaking, novel results of polarimetric millimeter wave (MMW) radars for automotive applications.

1.2 State of the Art and Literature Review

To the best of the author's knowledge, until today, only two dissertations exist that mainly focused on the topic of polarimetric radars for automotive applications. Both were published already two decades ago.
The first thesis was authored by Robert Schneider and emerged from results gathered at the Daimler research site in Ulm, Germany. It was published in May, 1998 [Sch98]. In this first dissertation a - to that date - highly advanced experimental fully polarimetric pulse radar system operating at 77 GHz with a range resolution of 1 m and with orthogonal linear polarizations was built out of single laboratory equipment devices (network analyzer, frequency synthesizer, etc.). In consequence, the sophisticated setup was extremely expensive, heavy, bulky, and thus served more as a proof of concept prototype than as a device that could be built into automobiles. Nevertheless, novel results respective to the influence of road and road side surfaces (asphalt, gravel, etc.) on the radar measurements were gained. Also, an analysis of multipath reflec-

tions over road surfaces and the associated interference effects was conducted. It was found, that the specific interference pattern that arises over distance for a point target, could be used to estimate its height. Concerning polarimetry, it was stated that in automotive radar measurements the aspect angle changes rapidly contrary to earth-observing SAR measurements, where polarimetry had already been firmly established. Therefore, only one attempt was made to find the amount of information contained in the fully polarimetric scattering matrix (S-matrix) by considering the cross-polarized (x-pol) to co-polarized (co-pol) radar cross section (RCS) ratio. It was found that by means of this ratio different targets (e.g. backside of truck and car) could be categorized, but that the same could be achieved by utilizing only the co-pol RCS. Based on this result, it was concluded that the polarimetric information would not be profitable and therefore not worth to be gathered due to the higher hardware effort needed for the setup. However, none of the numerous decompositions of the S-matrix that can be found in e.g. [LP09] were performed, nor other ways of extracting more polarimetric information from the S-matrix were analyzed.
The second thesis was published by Eric Shih-Syou Li, just a few months later than the first one, in October 1998, under [Shi98]. In this thesis, another advanced experimental radar system was assembled, consisting also of costly lab devices. The fully polarimetric radar system was operated at a higher frequency of 94 GHz, used circular polarization and a frequency-modulated continuous-wave (FMCW) modulation with a higher range resolution of 30 cm. Co-pol and x-pol radar images were generated in order to find the optimal antenna polarization combination (e.g. horizontal-horizontal (HH)) for a specific use case (e.g. mitigation of ghost target echoes). A detailed analysis of road surface/faults, road side objects and debris was conducted, in which theoretical models were proven by measurement results.

Both theses dealt partly with polarimetry and came to agreeing results, but also lack a crucial part of polarimetry - the evaluation of the complete complex S-matrix including the important relative polarimetric phase differences on which a part of the canonical object classification relies. By only evaluating the magnitudes of the co-pol and x-pol RCS values, this significant information remained unused.

More recently, four scientific papers, not related to the previously mentioned dissertations, were published. Two of them [Tru+16; Tru+17] present a fully polarimetric radar sensor operating at 77 GHz using orthogonal circular polarizations. The sensor's waveguide antenna comprises integrated corrugated horns and septum polarizers. Six transmitter (Tx) (operated in time-division

multiplexing (TDM) multiple-input, multiple-output (MIMO) mode) and 32 receiver (Rx) channels result in a high number of 192 virtual MIMO channel combinations, allowing a high angular resolution. However, both papers rather deal with the x-pol discrimination of the presented waveguide antenna, and similar to the afore-mentioned dissertations, only show the measured co-pol and x-pol RCS magnitudes of scattering centers on vehicles. The third paper [Erh+16] presents a modified automotive radar module using orthogonal linearly and circularly polarized patch antennas at 77 GHz. By reference to the change of the direction of rotation of the received circular polarizations, a dihedral reflector could be correctly classified versus a plate reflector. In the last paper [IBW17], orthogonal linearly polarized sectoral horn antennas are used in an experimental setup to record rotational measurements of traffic signs on a turntable. Each polarization combination is measured with only one Tx and one Rx horn at 77 GHz, respectively. Thus, no exploitable spatial information is available. The presented RCS measurement results show that the major contributions to the RCS are co-pol magnitudes whereas the x-pol RCS remains at low levels for all presented traffic signs.
In all papers, as in the dissertations, no polarimetric decompositions were investigated at all. Furthermore, no advanced classification algorithms were examined concerning target classification. Most of the reviewed works even lack a fundamental polarimetric calibration method. This thesis is thus the first work presenting a rigorous analysis and thorough investigation of radar polarimetry for automotive applications including innovative MIMO waveguide antenna designs, examination of feasible calibration methods, application of polarimetric decomposition algorithms, and classification of real traffic targets using modern deep learning approaches. Thereby, the thesis goes far beyond the simple evaluation of (uncalibrated) co-pol and x-pol RCS magnitude values, and dives deeper into the topic.

1.3 Outline of Thesis

The structure of the thesis is as follows:

- In Chapter 2, theoretical principles are introduced and the important concept of canonical objects is described. Furthermore, the polarimetric decompositions utilized in this thesis are explained and examples of polarimetric effects are given.

- Chapter 3 describes the experimental fully polarimetric radar system, and two MIMO waveguide antenna designs, using orthogonal linearly polarized sectoral horn antennas. Furthermore, one antenna is simulated and characterized in an anechoic chamber, and the results are compared.
- Chapter 4 deals with polarimetric calibration methods. A calibration method from [SUT90] is modified and generalized to the use of bistatic MIMO antennas. Calibration results of the experimental radar system are presented and the calibration quality is verified using the statistical evaluation of a large number of diverse canonical object measurements.
- In Chapter 5, fixed sensor measurements of rotated two-wheeled vehicles in an anechoic chamber are presented. The coherent polarimetric information is used to generate colorful images instead of only black and white radar images of the targets. Additionally, a theoretical and experimental analysis of polarimetric methods concerning the mitigation of multipath effects, including ghost targets, is conducted.
- In the second last Chapter 6, six artificial targets are constructed, using a varying number of different canonical objects mounted on a metal frame. The rotational signatures of the targets are measured and a deep learning approach is applied by training a convolutional neural network (CNN) in order to successfully classify the targets.
- Finally, the last Chapter 7 presents vehicular-based measurements of seven stationary real-world traffic targets, including among others, a car and a motorbike. The measurement dataset, that had been recorded with the radar sensor mounted in front of a measurement vehicle on a closed test track, was rigorously analyzed using modern machine learning and statistical methods. As before, a CNN is applied to successfully classify the targets. A fair comparison (in terms of the number of hardware antenna channels) of single-polarized and fully polarimetric radar systems is conducted resulting in promising results concerning the fully polarimetric system.

2 Basics of Radar Polarimetry and MIMO Radar

In this chapter, the most important concepts and methods that are necessary to understand the main achievements in this thesis, are introduced. The interested reader is directed to the respective literature for more details about MIMO radar in [LS08] and radar polarimetry in [Clo09; LP09; Mot06]. Parts of this chapter were published in [3; 2; 4].

2.1 Plane Waves and Polarization

The concept of wave polarization is most simply described by considering plane waves. A schematic figure of a plane wave is shown in Fig. 2.1a. It has a propagation direction indicated by the wave vector $\vec{k}$ and a polarization direction, indicated by the electric field vector $\vec{E}$. Thus, the oscillation direction of the electric field indicates the polarization of the wave. This vector can be decomposed into vertically and horizontally polarized parts, denoted by the unit vectors $\vec{e}_v$ and $\vec{e}_h$ in Fig. 2.1a. In this thesis, the x-direction is considered as horizontally polarized and the y-direction as vertically polarized, as implied in Fig. 2.1b with a y-polarized plane wave. Generally, a wave propagating in z-direction from observation plane $z = z_0$ at time $t = 0$ is given by [LP09]:

$$\vec{E} = \begin{pmatrix} E_x(t) \\ E_y(t) \end{pmatrix} = \begin{pmatrix} E_{0x} e^{-j(k_0 z_0 - \omega t - \Phi_x)} \\ E_{0y} e^{-j(k_0 z_0 - \omega t - \Phi_y)} \end{pmatrix} , \tag{2.1}$$

where E_{0x} and E_{0y} are the electric field amplitudes in the horizontal resp. vertical direction, k_0 is the wave number in free space, $\omega = 2\pi f$ is the angular frequency corresponding to the frequency f, and Φ_x , Φ_y are the corresponding true time phase delays for the vertical and horizontal polarization directions. This description is one of the most simple solutions of the wave equation. In

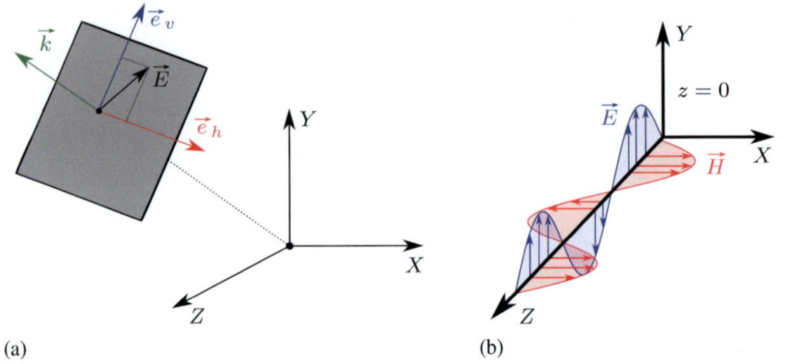

Figure 2.1: Illustrations of electromagnetic (EM) plane waves propagating in free space. (a): EM wave with planar wavefront, polarization direction and wave vector [Haj01]. (b): Y-polarized transverse electromagnetic (TEM) wave propagating in z-direction with electric and magnetic field vectors $\vec{E}$ and $\vec{H}$, respectively.

Eq. (2.1), the complex nature of the E-field vector is maintained and comprises amplitude as well as phase.

2.1.1 Jones Vector

In order to obtain a mathematical description of the coherent polarization state of an EM wave, the Jones calculus can be used. Based on Eq. (2.1), the Jones vector is denoted by [LP09]:

$$\vec{E}_J = \begin{pmatrix} E_{0x} e^{j\Phi_x} \\ E_{0y} e^{j\Phi_y} \end{pmatrix} . \tag{2.2}$$

According to the amplitudes and phase delays, all polarization states can be described by Jones vectors in a straightforward notation. For example, linear 45° polarization is described by the Jones vector $({}^{1}/_{\sqrt{2}}, {}^{1}/_{\sqrt{2}})^T$. For elliptical polarization with the special case of circular polarization, the phase delays have to be adjusted. For example, right-hand circular (RHC) polarization is described by $({}^{1}/_{\sqrt{2}}, {}^{-j}/_{\sqrt{2}})^T$ and left-hand circular (LHC) polarization by $({}^{1}/_{\sqrt{2}}, {}^{j}/_{\sqrt{2}})^T$. From various descriptions of the coherent polarization state e.g. the Poincaré sphere description, polarization ellipse and the Stokes vector description, which are presented in [LP09; Clo09; Mot06] and [Gol17], the one of the Jones vector is most convenient to introduce the S-matrix in the next section.

2.2 Polarimetric Scattering Matrix

The polarimetric S-matrix describes the connection between incident and scattered Jones vectors assuming a point scatterer in the far-field [Mot06]:

$$\vec{E}_{Js} = \frac{e^{-j\vec{k}\cdot\vec{r}}}{\sqrt{4\pi}|\vec{r}|}\begin{pmatrix} S_{11} & S_{12} \\ S_{21} & S_{22} \end{pmatrix}\vec{E}_{Ji} = \frac{e^{-j\vec{k}\cdot\vec{r}}}{\sqrt{4\pi}r}\boldsymbol{S}\vec{E}_{Ji}\,, \tag{2.3}$$

where $\vec{r}$ is the distance vector between the origin of the incident Jones vector $\vec{E}_{Ji}$ and the target, $\boldsymbol{S}$ is the S-matrix of the scatterer and $\vec{E}_{Js}$ is the scattered Jones vector at the location of the scatterer. The complex entries of the S-matrix can be defined in any orthogonal polarization basis like circular or linear polarization basis. These bases can be converted into another by matrix transformations that are given e.g. in [EA83]. In this thesis, without loss of generality, only linearly polarized radar systems are considered. Thus, the S-matrix can be formulated as:

$$\boldsymbol{S} = e^{j\phi_{hh}}\begin{pmatrix} |S_{hh}| & |S_{hv}|e^{j(\phi_{hv}-\phi_{hh})} \\ |S_{vh}|e^{j(\phi_{vh}-\phi_{hh})} & |S_{vv}|e^{j(\phi_{vv}-\phi_{hh})} \end{pmatrix}\,, \tag{2.4}$$

where the matrix entries are split into magnitude and phase. One of the absolute phase terms is taken out of the matrix and all other phases are represented as relative phases respective to this phase. Since the absolute phase is distance dependent and ambiguous if distances larger than one wavelength are considered, this phase has no physical meaning according to polarimetry. Therefore, only the relative phases of the matrix entries are evaluated. Generally, the S_{hv} component equals the S_{vh} component for monostatic radars due to reciprocity [LP09]. Consequently, the fully polarimetric S-matrix provides up to five times more information than its single-polarized counterpart (3 independent magnitudes and 2 relative phases) for monostatic radars. In case of a fully polarimetric bistatic radar, where the S_{hv} entry does not equal the S_{vh} entry, another independent magnitude and relative phase is added to the available

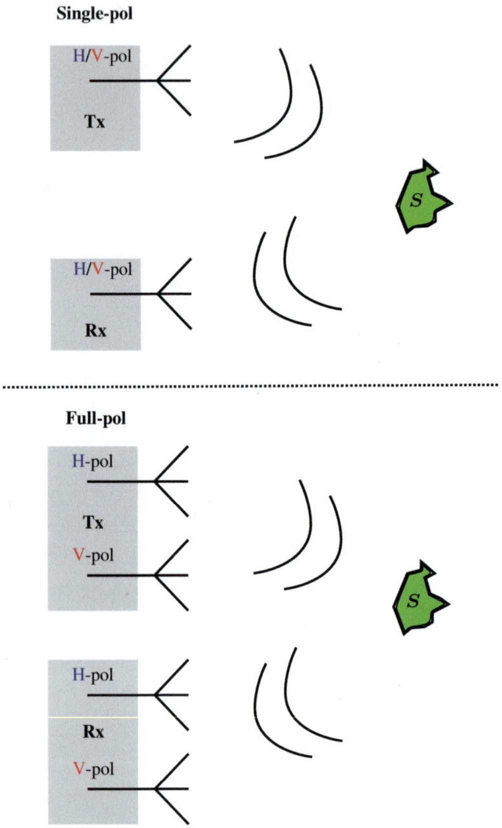

Figure 2.2: Comparison of single-polarized and fully polarimetric radar principles. Top: Example of linearly single-polarized radar principle with a backscatterer of scattering matrix $\boldsymbol{S}$. Bottom: Example of linearly polarized fully polarimetric radar principle with a backscatterer of scattering matrix $\boldsymbol{S}$.

information. In order to clarify this, a fully polarimetric radar system is compared to a single-polarized one in Fig. 2.2. According to the single-polarized radar system presented in the top of Fig. 2.2, the Tx and Rx antennas are co-pol in the same direction so that the matrix entries only comprise one co-pol entry:

$$\boldsymbol{S} = e^{j\phi_{hh}} \begin{pmatrix} |S_{hh}| & 0 \\ 0 & 0 \end{pmatrix} \quad \text{or} \quad \boldsymbol{S} = e^{j\phi_{vv}} \begin{pmatrix} 0 & 0 \\ 0 & |S_{vv}| \end{pmatrix}. \tag{2.5}$$

Since the absolute phase has no direct meaning as stated above, there is only one value in the corresponding matrix that carries information about the scatterer, the magnitude. In the case of a fully polarimetric radar system, schematically shown in the bottom of Fig. 2.2, the full S-matrix is attained. Two conditions must be met to achieve the complete measurement. First, at least two Tx channels with orthogonally polarized antennas must be operated orthogonally i.e. in a multiplexed procedure as e.g. TDM, frequency-division multiplex (FDM) or orthogonal frequency-division multiplexing (OFDM). Second, at least two orthogonally polarized Rx channels must be operated simultaneously so that the Rx signals are sampled and digitized at the same time. In this ideal example, the antennas have an infinite polarization isolation. In practice, the polarization purity strongly depends on the type of antenna element used and of mutual coupling and cross-talk between the channels. With an adequate antenna design though, high polarization isolations in the order of 30 dB and more can be achieved [Bal05; SUT90].

2.3 MIMO Radar

The concept of MIMO radar has evolved to a state-of-the-art-technology during the last decade [LS08]. Many of the principles have been adopted from communication theory [Bli08; Bli+09]. In order to demonstrate the MIMO radar principle, an exemplary schematic antenna system is presented in Fig. 2.3. The given system in Fig. 2.3a leads to a virtual array (VA) system in Fig. 2.3b if the transmitters are operated in an orthogonal manner, as discussed in Section 2.2. The resulting VA Rx-Tx combinations are given by a convolution of the Tx and Rx position vectors $\vec{r}_{tx}$ and $\vec{r}_{rx}$, respectively [LS08]. Virtually, the antenna aperture can be increased or gaps in the array, originating e.g. from physical antenna dimension restrictions, can be filled. Also, the virtual antenna positions can be designed to overlap, as is important e.g. for extraction of the polarimetric phase between two polarimetric chan-

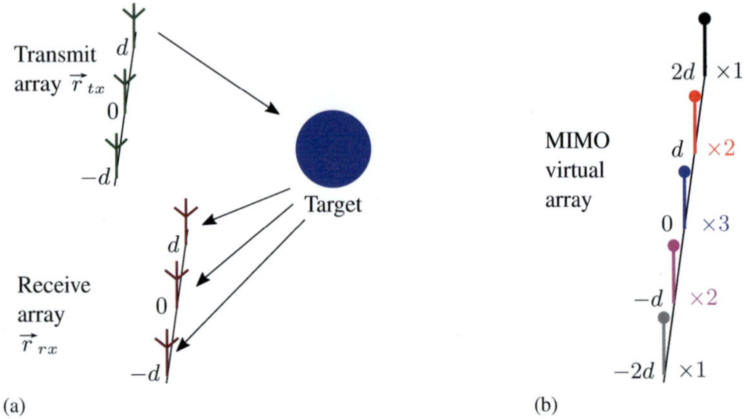

Figure 2.3: Principle of TDM MIMO radar from [BFF06] with three Tx and Rx antennas shown in (a), resulting in nine Rx-Tx combinations (partly overlapping), shown in (b).

nel combinations, independent of the angle of arrival (AOA) of the measured echo signal. What is measured with a radar and thus with the Rx-Tx combinations is the radar channel matrix that lies between transmitter and receiver. The entries of this channel matrix are given as phase term convolutions of the corresponding m^{th} Rx and n^{th} Tx antenna element positions [BFF06]:

$$H_{mn} \propto e^{j\vec{k}\cdot(\vec{r}_{tx,n}+\vec{r}_{rx,m})} \ . \tag{2.6}$$

For the example in Fig. 2.3, the channel matrix can be written as:

$$\boldsymbol{H} \propto \begin{pmatrix} e^{j\eta 2d} & e^{j\eta d} & e^{j\eta 0} \\ e^{j\eta d} & e^{j\eta 0} & e^{-j\eta d} \\ e^{j\eta 0} & e^{-j\eta d} & e^{-j\eta 2d} \end{pmatrix} , \tag{2.7}$$

where $\eta = \vec{k} \cdot \vec{e}_d$ with $\vec{e}_d$ being the unit vector in direction of the Rx array elements. The colors emphasize the matrix entries matching with the resulting VA in Fig. 2.3b.

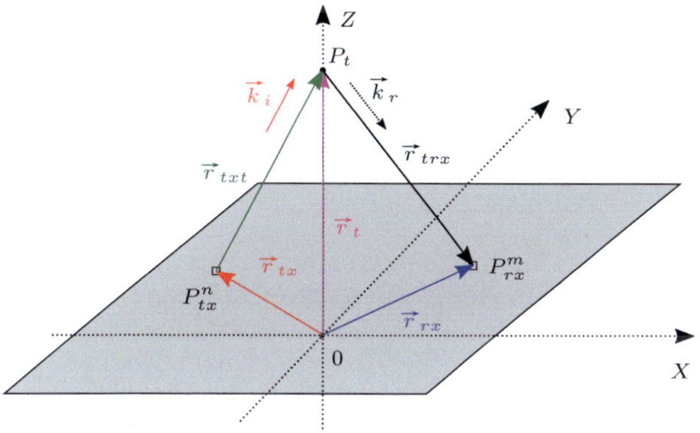

Figure 2.4: Backscattering scenario coordinate system for a point target at boresight direction.

2.3.1 Calculation of Virtual Array Positions

In order to derive the result of Eq. (2.6), the MIMO scattering scenario shown in Fig. 2.4 is considered. It shows the aperture of the MIMO-antenna with a point target in boresight direction at position P_t in a bistatic configuration. In the following, plane waves are considered, the backscatter alignment (BSA) convention from [LP09] is used and the harmonic time dependence factor $e^{j\omega t}$ from Eq. (2.1) is suppressed. The physical positions of the phase centers of the nth Tx- and m^{th} Rx-antennas on the aperture are denoted as P^n_{tx} and P^m_{rx} with their corresponding position vectors $\vec{r}_{tx}$ and $\vec{r}_{rx}$, respectively. Pointing from the nth Tx-antenna position to the target, the vector $\vec{r}_{txt}$ shows the path on which the transmitted signal travels from the nth Tx to the target. Similarly, the vector $\vec{r}_{trx}$ denotes the path that the scattered signal travels from the target to the mth Rx-antenna. In the following, far-field conditions are considered. Therefore, two assumptions are made: $r_t >> (r_{rx}, r_{tx})$ and $\vec{k}_i \approx -\vec{k}_r$. The reflected electric field at the mth Rx-position is connected to the incident electric field from the nth Tx at the target by the mn^{th} polarization combination component of the S-matrix $\boldsymbol{S}$ by [Mot06]:

$$E^r_{mn} = \frac{S_{mn}}{\sqrt{4\pi} r_{trx}} E^i_n e^{-j\vec{k}_r \cdot \vec{r}_{trx}} \, . \tag{2.8}$$

For the incident electric field at the target, a plane wave coming from the nth Tx-position is formulated: $E_n^i = E_n^t e^{-j\vec{k}_i \cdot \vec{r}_{txt}}$. Thus, the ratio of the received signal at the mth Rx from the nth Tx is given by:

$$\frac{E_{mn}^r}{E_n^t} = \frac{S_{mn}}{\sqrt{4\pi} r_{trx}} e^{-j\vec{k}_i \cdot [2\vec{r}_t - (\vec{r}_{tx} + \vec{r}_{rx})]} . \tag{2.9}$$

This result shows that the components of the channel matrix $\boldsymbol{H}$ which connect the nth Tx- to the mth Rx-signal are proportional to the convolution of phase terms from the phase center positions of Tx and Rx as in Eq. (2.6) and [LS08]:

$$H_{mn} = E_{mn}^r / E_n^t \propto e^{j\vec{k}_i \cdot \vec{r}_{tx}} * e^{j\vec{k}_i \cdot \vec{r}_{rx}} . \tag{2.10}$$

2.3.2 DOA estimation for an 1D-ULA

The principle of direction of arrival (DOA) estimation for an 1D-uniform linear array (ULA) with N receiving elements is demonstrated by an off-broadside impinging plane wave in Fig. 2.5. The electric field of the received

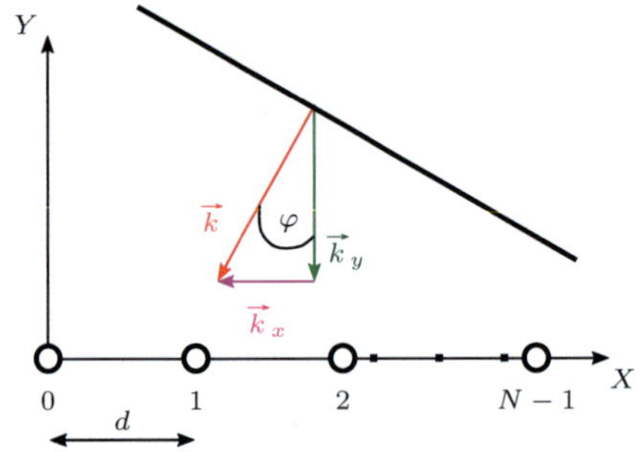

Figure 2.5: 1D-ULA and off-broadside impinging plane wave at an angle φ.

plane wave at the nth Rx antenna element is described by:

$$E_{rn} = E_{0r} e^{-j\vec{k} \cdot \vec{r}_n} = E_{0r} e^{-j(\vec{k}_x + \vec{k}_y) \cdot nd\vec{e}_x} \tag{2.11}$$

$$E_{rn} = E_{0r} e^{j\frac{2\pi}{\lambda} \sin(\varphi) nd} , \; n \in [0, N-1] , \tag{2.12}$$

where $\vec{r}_n$ is the position vector of the nth antenna element and λ the free space wavelength. The Nyquist-Shannon theorem states that an azimuthal AOA φ can be reconstructed without error, if the plane wave is spatially sampled with the double spatial frequency ν [Hua+11]: $\nu_s \geq 2\nu = 2/\lambda$. Thus, the minimum distance of Rx antenna elements must be $d \leq 1/\nu_s = \lambda/2$, which corresponds to a $180°$ phase shift. An array with this spacing between its elements is called $\lambda/2$ ULA. If the spacing becomes larger than $\lambda/2$, grating lobes emerge which cause ambiguities in the spatial spectrum [Hua+11]. In this thesis, quasi $\lambda/2$ ULAs are used for the DOA estimation. Thus, the spatial frequency can be simply estimated by a fast Fourier transform (FFT) processing and peak extraction. The estimated AOA from Eq. (2.12) is then:

$$\varphi_{est} = \arcsin\left(\lambda \nu_{est}\right), \tag{2.13}$$

where ν_{est} is the estimated spatial frequency.

2.4 Speckle, Glint & Scintillation

Destructive interference-based effects can lead to a deterioration of the radar system performance and are inherently given by the underlying physics of the scatterers. Most of these effects originate from interfering backscattered waves by scattering centers of the targets, and interaction with the limited resolution of the radar. From [And+13; SBW08; Bha+98] it is known that targets of any size can mainly be described by a certain number of scattering centers, depending mainly on the measurement frequency and bandwidth. The RCS, size and number of these scattering centers depend on the measurement wavelength and spatial resolution of the radar which is given by the size of the (virtual) antenna aperture and the measurement bandwidth. One phenomenon that occurs if the target's scattering centers are smaller than the resolution cells of the radar is called speckle. The principle of speckle is shown in Fig. 2.6 for an airborne SAR sensor. Many elementary scattering centers lie in single resolution cells. The measured RCS is then a combination of the RCS contributions of all these scattering centers. The coherent superposition can result in a rapid change of the total field vector between adjacent resolution cells. Due to the varying positions of the elementary scattering centers, speckle is highly dependent on the aspect angle under which the target is observed. An example of the resulting total scattering field vectors of two adjacent resolution cells is given in Fig. 2.7. Speckle influences amplitude and phase of the total field vector which means that RCS and polarimetric phase contributions can be affected. Nevertheless, the effect is deterministic, i.e. seen under the same aspect angle, a target provides the same polarimetric RCS. It is expected that targets which are large compared to the measurement wavelength and have a complex geometry like e.g. vehicles in traffic give rise to speckle and thus highly fluctuating RCS over aspect angle for automotive radars operating at MMW ranges. This phenomenon can also be observed in the field of air surveillance radar where the observed aircrafts are much bigger than the measurement wavelength [Kno12]. In diverse literature sources, the effect of the target's RCS fluctuations over aspect angle is also called glint or scintillation [Sko02; Avi13]. Basically, the same coherent fluctuations are meant that result in moving apparent scattering centers of large targets over aspect angle, where the elementary single scattering centers cannot be resolved sufficiently by the radar. The "random" movement is caused by interference of the backscattered waves by the scattering centers which are seen at a specific aspect angle that is dependent of the motion and/or vibration of the radar and/or target itself.

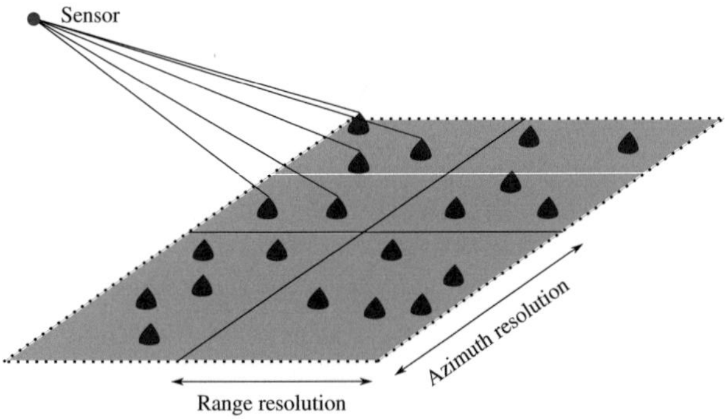

Figure 2.6: Principle of speckle caused by different elementary scattering contributions in each resolution cell for an airborne SAR sensor [Haj01].

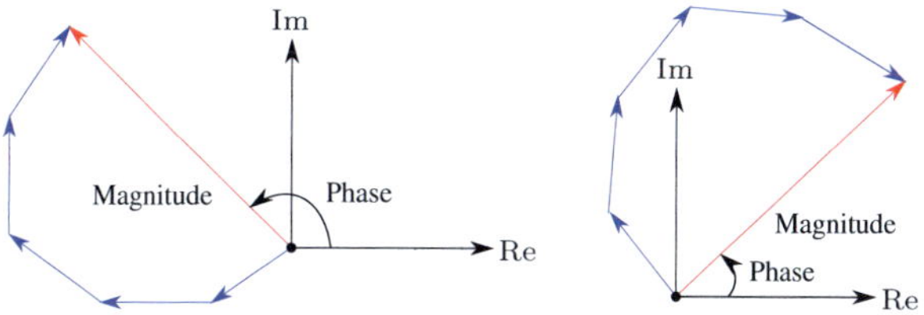

Figure 2.7: Principle of speckle as a coherent superposition of complex scattering parts in one resolution cell, shown for two adjacent resolution cells [Haj01].

2.5 Canonical Objects

In radar polarimetry, four basic physical scattering phenomena exist. They can be described by so-called canonical objects. Any existing S-matrix can be composed as a weighted superposition of these basic scattering principles. Primarily, canonical objects can be characterized by co-pol or x-pol scattering. For co-pol scatterers, a further selection by the parity of reflections, i.e. by an even or odd number of bounces, can be performed. Furthermore, polarizing behavior can be detected. In Fig. 2.8 a list of ideal canonical objects (perfectly conducting material, no edge scattering) is presented, each with incoming and reflected horizontal and vertical electric fields. In Figs. 2.8a, 2.8b and 2.8e, co-pol canonical objects with an odd number of bounces (reflections) are illustrated, whereas in Fig. 2.8c, a co-pol even bounce canonical object is presented. Furthermore, in Fig. 2.8d, a x-pol canonical object is given, whereas in Fig. 2.8f an example of a polarizing object as a wire grid is shown. All of

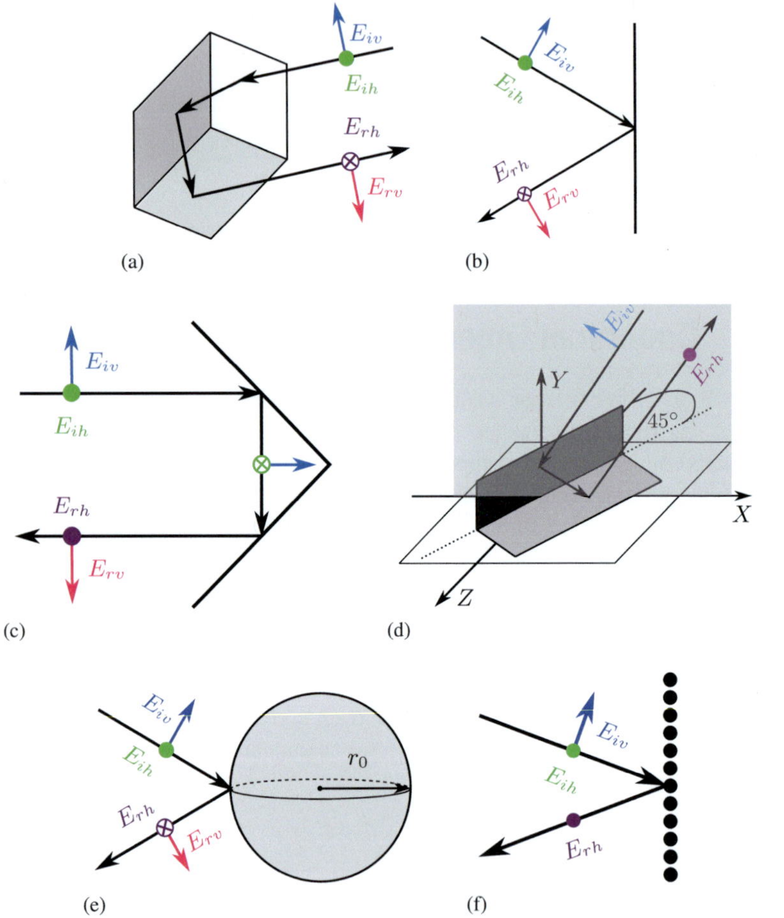

Figure 2.8: List of canonical objects with incoming and reflected electric fields. (a): Trihedral corner reflector with co-polar triple bounce reflections. (b): Plate reflector with co-polar single bounce reflection. (c): Horizontally oriented dihedral corner reflector with co-polar double bounce reflections. (d): $45°$ oriented dihedral corner reflector with cross-polar double bounce reflections. (e): Sphere reflector with co-polar single bounce reflections. (f): Horizontally oriented wire-grid polarizer with horizontally polarized single bounce reflection.

the four different presented kinds of canonical objects can be described by a characteristic S-matrix. These ideal S-matrices are given by [EA83]:

$$\boldsymbol{S_{cpob}} = \sqrt{\sigma_{cpob}} \begin{pmatrix} 1 & 0 \\ 0 & 1 \end{pmatrix} \tag{2.14a}$$

$$\boldsymbol{S_{cpeb}} = \sqrt{\sigma_{cpeb}} \begin{pmatrix} 1 & 0 \\ 0 & -1 \end{pmatrix} \tag{2.14b}$$

$$\boldsymbol{S_{xp}} = \sqrt{\sigma_{xp}} \begin{pmatrix} 0 & 1 \\ 1 & 0 \end{pmatrix} \tag{2.14c}$$

$$\boldsymbol{S_{p}} = \sqrt{\sigma_{p}} \begin{pmatrix} 1 & 0 \\ 0 & 0 \end{pmatrix} \vee \sqrt{\sigma_{p}} \begin{pmatrix} 0 & 0 \\ 0 & 1 \end{pmatrix} , \tag{2.14d}$$

where σ_{cpob}, σ_{cpeb}, σ_{xp} and σ_p are the corresponding RCS values of the co-pol odd bounce, co-pol even bounce, x-pol and polarizing objects, respectively. The S-matrix for the wire grid polarizer in Eq. (2.14d) is given exemplary for a perfect horizontal polarizer by the left S-matrix, and for an ideal vertical polarizer by the right one.

A very versatile canonical object is the dihedral, shown in Figs. 2.8c and 2.8d and more specifically in Fig. 2.9a. It consists of two perfect electrically conducting (PEC) plates with width w and height h that are arranged in a perpendicular orientation to each other. The polarimetric backscattering characteristics of the dihedral change depending on the rotation around its seam. An ideal dihedral object can be described for arbitrary rotation tilt angles θ by the following S-matrix [EA83; Mot06]:

$$\boldsymbol{S_{d}} = \sqrt{\sigma_{d}} \begin{pmatrix} \cos(2\theta) & \sin(2\theta) \\ \sin(2\theta) & -\cos(2\theta) \end{pmatrix} , \tag{2.15}$$

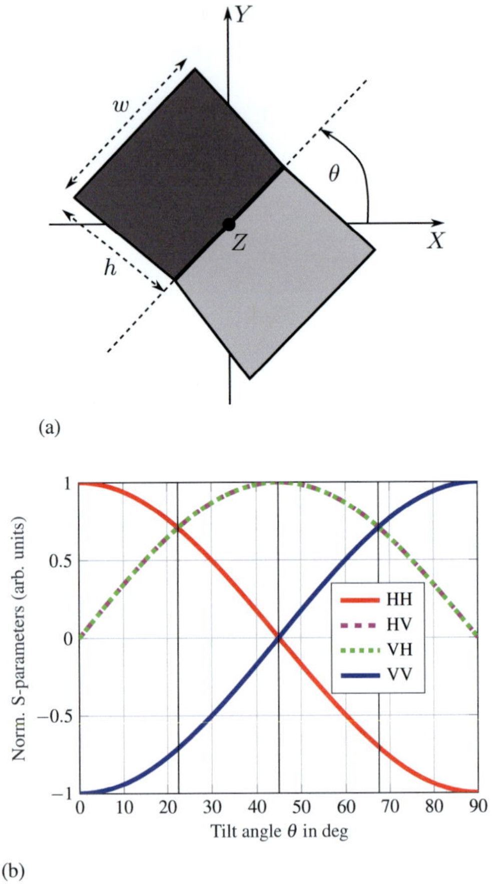

Figure 2.9: (a): View of dihedral rotated by an angle θ around its seam. (b): Normalized S-parameter graphs of the dihedral over rotation tilt angle.

where σ_d is the RCS of the dihedral. This scattering behavior is visualized by the graphs in Fig. 2.9b. Three points on these graphs are of special interest. For example, if orientated in a 0° or 90° rotation (horizontally or vertically orientated), the dihedral represents a co-pol even bounce object, whereas if orientated in a 45° rotation, it represents a pure x-pol object. It can be observed, that in rotation angles of $22.5°$ and $67.5°$ the dihedral can also represent a mixed target with a scattering matrix of:

$$\boldsymbol{S}_{mt} = \sqrt{\frac{\sigma_d}{2}} \begin{pmatrix} 1 & 1 \\ 1 & -1 \end{pmatrix} , \tag{2.16}$$

which is a superposition of the co-pol even bounce and x-pol canonical objects.

2.5.1 Canonical Objects in Real Environments

After introduction of the ideal canonical objects in the foregoing section, the question arises how these canonical objects occur in practical sceneries. The connection between real physical scatterers with complex shapes and large sizes, and ideal canonical objects is not directly obvious.
In SAR polarimetry, the canonical scattering mechanisms occurring in real environments are more obvious. In comparison to the corresponding mechanisms in the automotive field, a list of three basic mechanisms is given schematically in Fig. 2.10. In Fig. 2.10a, co-pol odd bounce surface scattering is shown, in SAR polarimetry occurring e.g. in plane landscapes and ocean surfaces. The corresponding phenomenon in the automotive case is shown in Fig. 2.10b where the surface scattering comes from the metallic surfaces at the back of vehicles. The second canonical phenomenon which can be observed is the co-pol even bounce scattering, occurring in SAR polarimetry e.g. at the fronts of high buildings. This is shown in Fig. 2.10c, whereas the corresponding scattering scenario in the automotive field is presented in Fig. 2.10d for an even bounce reflection at a curbstone. Finally, a mechanism that was principally not introduced as a canonical scattering mechanism yet, is volume scattering. Detection is possible, if several S-matrices are averaged, e.g. in SAR polarimetry between several pixels of the SAR image. Volume scattering can then be determined by a measure of the depolarization of the incoming strongly polarized waves. An introduction to depolarization is presented together with an effective method of detecting volume scattering in the following Section 2.6.2. In real environments, volume scattering can be observed for example by the backscatter from a tree, as shown schematically in Fig. 2.10e. Analogously, in the field of automotive polarimetry, volume scattering occurs at road side vegetation or grass, as indicated in Fig. 2.10f.

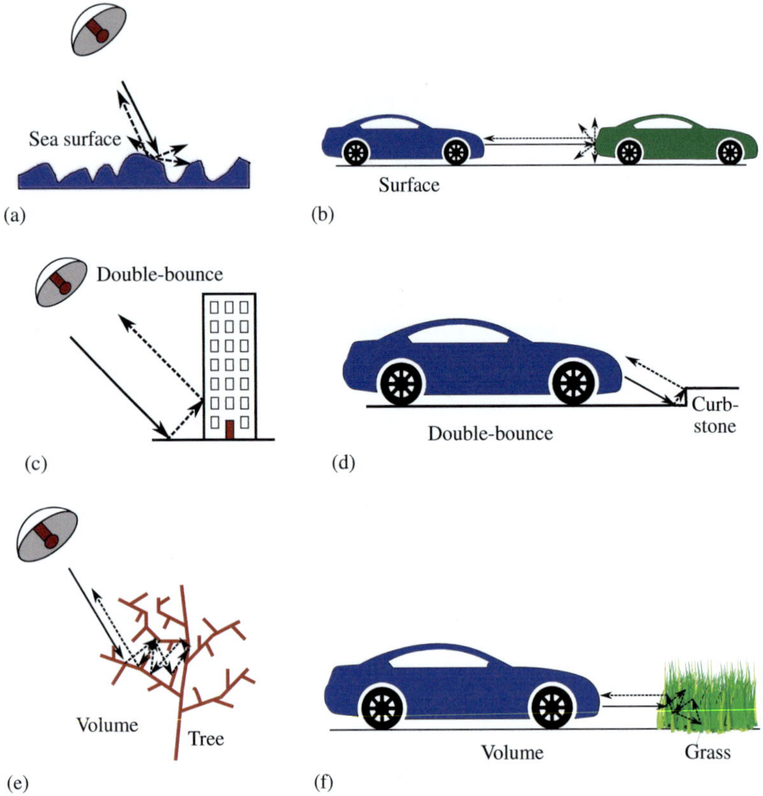

Figure 2.10: (a), (c), (e): Schematic view of SAR-based scattering phenomena. (b), (d), (f): Corresponding automotive radar-based scattering phenomena.

Another good example of canonical scattering mechanisms occurring especially in MMW radar measurements, is edge scattering. It originates from metallic edges which are thin compared to the wavelength. The principle is demonstrated schematically in Fig. 2.11 for a thin metallic edge with incoming and reflected electric fields. In analogy to a thin wire or dipole antenna, the backscattered radiation is strongly polarized in the direction parallel to the edge [Bal05; Kra88]. Complementary, for a thin metallic slot, e.g. if two thin edges are facing each other, as schematically shown in Fig. 2.12, the polarization direction of the backscattered fields is orthogonal to the one in case of

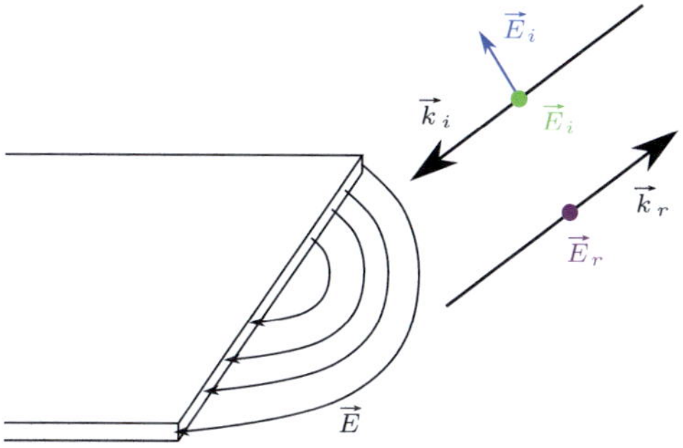

Figure 2.11: Incoming and reflected electric fields at a thin metallic edge which acts as a polarizer parallel to the edge, according to the left S-matrix from Eq. (2.14d).

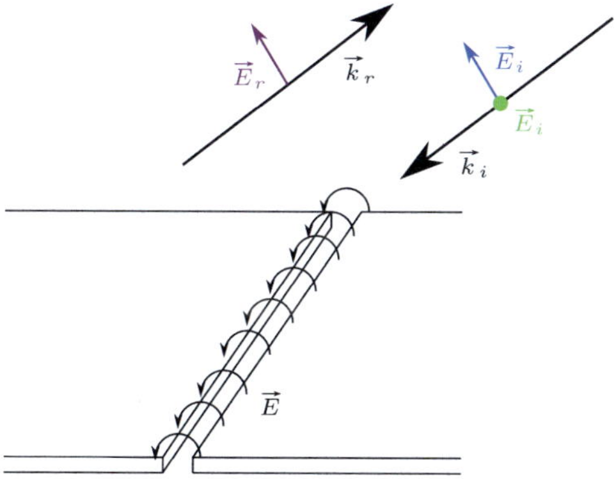

Figure 2.12: Incoming and reflected electric fields at a thin metallic slot which acts as a polarizer perpendicularly to the slot, according to the right S-matrix from Eq. (2.14d).

the metallic edge. Also in this case, the analogy of the reflected fields to the radiation of a thin slot antenna is given [Bal05; Kra88]. The reflected fields are mostly polarized across the slot direction, as indicated in Fig. 2.12.

2.6 Polarimetric Decompositions

Subsequent to having measured the polarimetric S-matrix, several signal processing approaches exist. The intention followed by these approaches is to extract the polarimetric information in a physical meaningful way. Therefore, the S-matrix needs to be decomposed. In the following, two important methods of matrix decomposition are discussed that are used in this thesis.

2.6.1 Coherent Decompositions

Direct decompositions of a specific S-matrix are called coherent. Coherency in the sense of decompositions means that one single S-matrix is considered for the characterization of each resolution cell of the radar without any further averaging between the cells. Any given S-matrix can be decomposed into a coherent sum of basis matrices, each weighted by coefficients. These matrices and their coefficients correspond to canonical scattering mechanisms. Numerous 2×2 matrix decompositions exist and are discussed in literature, e.g. [LP09; Mot06; Clo09; Alb+04; Boe07]. All of these decompositions lead to interpretable results in polarimetry - but not all of these results are also realized in nature which means that they partly lack a connection to the before introduced canonical objects in Section 2.5. Therefore, only the most intuitive coherent decomposition is presented here exemplarily.

Pauli Decomposition

One of the most common coherent decompositions is the Pauli decomposition, its name originated from the Pauli matrices, which are well-known, e.g. from quantum mechanics. The Pauli matrices form a set of four complex unitary and Hermitian matrices and build a basis for the vector space of 2×2 Hermitian matrices. These matrices are given as [FLS66]:

$$\boldsymbol{\sigma_0} = \boldsymbol{I} = \begin{pmatrix} 1 & 0 \\ 0 & 1 \end{pmatrix}, \boldsymbol{\sigma_1} = \begin{pmatrix} 0 & 1 \\ 1 & 0 \end{pmatrix}, \\ \boldsymbol{\sigma_2} = \begin{pmatrix} 0 & -j \\ j & 0 \end{pmatrix}, \boldsymbol{\sigma_3} = \begin{pmatrix} 1 & 0 \\ 0 & -1 \end{pmatrix}. \tag{2.17}$$

They are used as basis for the polarimetric Pauli decomposition which is written as [LP09]:

$$\boldsymbol{S} = \frac{a}{\sqrt{2}}\boldsymbol{\sigma_0} + \frac{b}{\sqrt{2}}\boldsymbol{\sigma_3} + \frac{c}{\sqrt{2}}\boldsymbol{\sigma_1} + \frac{d}{\sqrt{2}}\boldsymbol{\sigma_2}\,. \tag{2.18}$$

The complex coefficients a, b, c and d can be calculated by simple coherent additions and subtractions of the S-matrix components:

$$\begin{aligned} a &= \frac{S_{hh} + S_{vv}}{\sqrt{2}},\ b = \frac{S_{hh} - S_{vv}}{\sqrt{2}}, \\ c &= \frac{S_{hv} + S_{vh}}{\sqrt{2}},\ d = j\frac{S_{hv} - S_{vh}}{\sqrt{2}}\,. \end{aligned} \tag{2.19}$$

Each of the Pauli coefficients is related to a physical canonical scattering process. Therefore, the squared absolute value of each Pauli component describes that part of the complete RCS which is stored in the specific corresponding canonical scattering process. These canonical scattering processes are coded in the Pauli components:

- The RCS $|a|^2/2$ is related to odd bounce scattering such as scattering from flat surfaces, trihedral corner reflectors and spheres.
- The RCS $|b|^2/2$ is related to even bounce scattering, given by e.g. dihedral corner reflectors rotated by $0°$ and $90°$, respectively.
- The RCS $|c|^2/2$ is related to cross-polar scattering phenomena which occur in dipole and dihedral scattering, if the objects are rotated by $45°$ around the radar line-of-sight (LOS).
- The RCS $|d|^2/2$ describes the entire asymmetric RCS that is stored in the decomposed S-matrix. For monostatic systems this RCS is theoretically zero because of the reciprocity theorem ($S_{hv} = S_{vh}$) [LP09]. Nevertheless, for the real quasi-monostatic system used in this work, a small fraction of scattered power is still stored in the RCS of this component. It can be used to obtain an average 4×4 coherency or covariance matrix which is based on averaged data and second order statistics [LP09]. This topic is dealt with later on in Section 2.6.2.

From Eq. (2.14) and Eqs. (2.18) and (2.19) it can be derived that the total polarimetric RCS σ_{pol} of the S-matrix is calculated by the trace (sum of the main diagonal elements) of $\boldsymbol{S}\boldsymbol{S}^{\dagger}$:

$$\sigma_{pol} = \frac{1}{2}\,\mathrm{tr}\left(\boldsymbol{S}\boldsymbol{S}^{\dagger}\right) = \frac{1}{2}\left(|a|^2 + |b|^2 + |c|^2 + |d|^2\right) \,. \tag{2.20}$$

Finally, the general bistatic Pauli scattering vector combines all Pauli parameters [Clo09]:

$$\mathbf{p} = (a, b, c, d)^T \,, \tag{2.21}$$

where in the strict monostatic case the fourth component can be neglected, since then $S_{hv} = S_{vh}$ due to reciprocity, so that d from Eq. (2.19) equals zero.

2.6.2 Incoherent Decompositions

After having understood how to extract the basic canonical scattering phenomena from the S-matrix, the next step is to develop a method how to acquire the remaining polarimetric information which is not directly related to one single S-matrix. Rather, only averaging of several S-matrices can reveal the additional related polarimetric effects. In the following, the phenomenon of depolarization (not to be confused with cross-polarization) is introduced. Furthermore, the concept of incoherent decompositions which includes any sort of averaging single S-matrices, is described.

Depolarization Effects

As mentioned in Section 2.5.1, information that becomes available after averaging is related to depolarizing effects and occurs in e.g. volume scattering. For example, Fig. 2.10e or Fig. 2.10f is considered for an incoming wave. Let the time of flight (TOF) from the moment a wave emanates from the radar transmitter until the backscattered wave impinges on the receiver be much shorter than the spatial changes of the target and radar sensor over time. Many bounces of the incoming wave on randomly aligned branches, tree leaves or grass stalks lead to a randomly polarized outgoing wave. This backscattered wave is arbitrarily polarized depending on the aspect angle. Due to elapsing time, the position of radar and/or target changes, e.g. the grass stalks are moving in the wind. A second wave sent out to the target at another moment of time results in a second backscattered wave with a deviating polarization to the first one. This means that the polarization of the target has become time and/or space dependent. In such a case, it is not sufficient to focus on just one measured S-matrix, since it has become time-variant. Such processes are called to be depolarizing, as they are not deterministic anymore but instead underly a statistical nature over time and/or spatial changes [Clo09]. Obviously, in order to unveil this nature, also statistical methods play a significant role. Because it changes the polarization state of the incoming waves in an unpredictable way, so that the "polarization coherency" is destroyed, depolarization is always connected to a loss of information. Depolarizing targets are also labeled "distributed targets" or "partial scatterers" in the respective literature like e.g. [Clo09; Mot06; LP09].

Coherency Matrix

Following this idea, the concept of the so-called coherency matrix $\boldsymbol{T}$ was developed, which is based on the outer product of the Pauli scattering vector from Eq. (2.21) [Clo09]:

$$\boldsymbol{T} = \mathbf{p}\mathbf{p}^{\dagger} = \frac{1}{2}\begin{pmatrix} |a|^2 & ab^* & ac^* & ad^* \\ ba^* & |b|^2 & bc^* & bd^* \\ ca^* & cb^* & |c|^2 & cd^* \\ da^* & db^* & dc^* & |d|^2 \end{pmatrix} = \boldsymbol{T}^{\dagger}\,. \tag{2.22}$$

A similar matrix called the covariance matrix $\boldsymbol{C}$, is introduced in literature, as in e.g. [Clo09]. It can be simply converted into the coherency matrix $\boldsymbol{T}$ and contains the same information; it is thus not further discussed here.
As can be seen in Eq. (2.22), the coherency matrix is a square and positive semidefinite Hermitian matrix and has therefore important and significant features [Str05]:

- One is that the diagonal entries of $\boldsymbol{T}$ are all real and positive or zero - the trace results in the total polarimetric scattering power, as in Eq. (2.20).

- Another very important one is that $\boldsymbol{T}$ only has real eigenvalues which are greater or equal to zero.

- According to the spectral theorem of linear algebra, $\boldsymbol{T}$ is always diagonalizable by an unitary matrix $\boldsymbol{U}$ and the diagonalized matrix $\boldsymbol{D}$ has only real entries, where the elements on its main diagonal are the eigenvalues of $\boldsymbol{T}$.

- The columns of $\boldsymbol{U}$ are built by orthonormal eigenvectors of $\boldsymbol{T}$.

These features help to decompose $\boldsymbol{T}$ to extract and evaluate the statistical polarimetric information which is stored in its eigenvalues and eigenvectors. Using the spectral theorem from above, the diagonalized form of $\boldsymbol{T}$ can be written as:

$$\boldsymbol{T} = \boldsymbol{U}\boldsymbol{D}\boldsymbol{U}^{-1} = \boldsymbol{U} \begin{pmatrix} \lambda_1 & 0 & 0 & 0 \\ 0 & \lambda_2 & 0 & 0 \\ 0 & 0 & \lambda_3 & 0 \\ 0 & 0 & 0 & \lambda_4 \end{pmatrix} \boldsymbol{U}^{\dagger} , \tag{2.23}$$

with $\lambda_{1..4}$ being the eigenvalues of $\boldsymbol{T}$. Since no averaging was performed in Eq. (2.23), all eigenvalues except one are zero. This follows from the fact that $\boldsymbol{T}$ is then a singular matrix with rank one [Str05]. In order to obtain a positive non-degenerate eigenvalue spectrum, averaging must be applied:

$$\langle \boldsymbol{T} \rangle = \left\langle \mathbf{p}\mathbf{p}^{\dagger} \right\rangle = \begin{pmatrix} < |a|^2 > & < ab^* > & < ac^* > & < ad^* > \\ < ba^* > & < |b|^2 > & < bc^* > & < bd^* > \\ < ca^* > & < cb^* > & < |c|^2 > & < cd^* > \\ < da^* > & < db^* > & < dc^* > & < |d|^2 > \end{pmatrix} . \tag{2.24}$$

Averaging is arbitrary and can be done in any dimension that seems reasonable and underlies statistical changes. In SAR polarimetry, averaging over several pixels (resolution cells) is performed by moving average windows over the 2D SAR image [Alb04]. On the other hand, in the automotive radar field, SAR cannot be used in a beneficial way for the forward looking radar. Thus, also other averaging concepts come into consideration. Averaging over a period of time (which is closely related to tracking) or in different dimensions like range, velocity and AOA spectrum can be performed. However, any averaging

in the dimensions where time- or space-variant changes are assumed can be reasonable.
Subsequent to the performed averaging, the eigenvalue spectrum from Eq. (2.23) is filled with non-degenerate and real positive eigenvalues which have to be ordered in a descending order on the main diagonal of $\boldsymbol{D}$ ($\lambda_1 \geq \lambda_2 \geq \lambda_3 \geq \lambda 4$). The fourth eigenvalue in Eq. (2.23) can be considered as pure additive white Gaussian noise (AWGN) with noise power N in (quasi-) monostatic radar systems so that the eigenvalues can be written as [HPC02; HPC01]:

$$\lambda_1 = \lambda_1' + N \ , \ \lambda_2 = \lambda_2' + N \ , \ \lambda_3 = \lambda_3' + N \ , \text{ and } \lambda_4 = N \ . \tag{2.25}$$

If the fourth eigenvalue is subtracted from the first three eigenvalues, the AWGN can be removed so that a noise reduction is reached. With the resulting rank reduction of $\boldsymbol{D}$, the unitary matrix from Eq. (2.23) can now be written as [Clo09]:

$$\boldsymbol{U} = \begin{pmatrix} \cos\alpha_1 e^{j\epsilon_1} & \cos\alpha_2 e^{j\epsilon_2} & \cos\alpha_3 e^{j\epsilon_3} \\ \sin\alpha_1 \cos\beta_1 e^{j\delta_1} & \sin\alpha_2 \cos\beta_2 e^{j\delta_2} & \sin\alpha_3 \cos\beta_3 e^{j\delta_3} \\ \sin\alpha_1 \sin\beta_1 e^{j\gamma_1} & \sin\alpha_2 \sin\beta_2 e^{j\gamma_2} & \sin\alpha_3 \sin\beta_3 e^{j\gamma_3} \end{pmatrix} \tag{2.26a}$$

$$\boldsymbol{U} = \begin{pmatrix} e_{11} & e_{21} & e_{31} \\ e_{12} & e_{22} & e_{32} \\ e_{13} & e_{23} & e_{33} \end{pmatrix} = \begin{pmatrix} \mathbf{e}_1 & \mathbf{e}_2 & \mathbf{e}_3 \end{pmatrix} , \tag{2.26b}$$

where the columns of $\boldsymbol{U}$ correspond to the three eigenvectors $\mathbf{e}_{1..3}$ of $\boldsymbol{T}$ with each five degrees of freedom, expressed here in terms of angles. Since the complex argument phase angles are ambiguous (like the range dependent absolute phase angle in Eq. (2.4)) and thus not important for the following analysis, they are not further considered here. Rather, the α and β angles contribute to physically interpretable polarimetric mechanisms, as will be introduced in the following section.

Entropy / Alpha Angle (H / α) Decomposition

Returning to the noise reduced and ordered eigenvalue spectrum of Eq. (2.25), it also becomes apparent that the eigenvalues play an important role in the distribution of these mechanisms, as the total scattered power is distributed over the eigenvalues. From a statistical viewpoint, the definition of quantitative probabilities is feasible [Clo09]:

$$P_i = \frac{\lambda_i'}{\sum_{j=1}^{3} \lambda_j'} \; . \tag{2.27}$$

The idea of the eigenvalue probabilities can then be extended to the definition of an entropy value H that becomes a quantitative measure of the scattering polarization “randomness" [Clo09]:

$$H = -\sum_{i=1}^{3} P_i \log_3 (P_i) \; . \tag{2.28}$$

By this definition, the entropy is zero for a completely deterministic scattering process without loss of polarimetric information, whereas an entropy of one means that the scattering process is completely random and depolarizing, thus all polarimetric information is lost. In this case, the knowledge of the entropy being one is still an important information which tells that the process is completely depolarizing. Ranging from zero, where the scattering process can be perfectly decomposed into the canonical scattering parts (e.g. by the Pauli decomposition from Section 2.6.1), to one, where no classification by polarimetric means is possible anymore, there is a large range of entropy values in between zero and one. The properties of the partly depolarizing systems that lie in this range can still be split into canonical scattering phenomena and additionally into depolarizing characteristics. Apart from the depolarizing part, the deterministic part of information which contains the canonical scattering phenomena can be obtained by evaluating the eigenvectors from Eq. (2.26), where the single α and β angles are calculated by [Haj01]:

$$\alpha_1 = \arccos\left(|e_{11}|\right) \quad \alpha_2 = \arccos\left(|e_{21}|\right) \quad \alpha_3 = \arccos\left(|e_{31}|\right) \tag{2.29a}$$

$$\beta_1 = \arctan\left(\frac{|e_{13}|}{|e_{12}|}\right) \quad \beta_2 = \arctan\left(\frac{|e_{23}|}{|e_{22}|}\right) \quad \beta_3 = \arctan\left(\frac{|e_{33}|}{|e_{32}|}\right) \; . \tag{2.29b}$$

In terms of statistical maximum likelihood, the single angle values can be combined by summing up their probability weighted versions [CP97]:

$$\alpha = \sum_{i=1}^{3} P_i \alpha_i \quad \text{and} \quad \beta = \sum_{i=1}^{3} P_i \beta_i \ . \tag{2.30}$$

The α angle can then be best described by considering Fig. 2.13. It can be

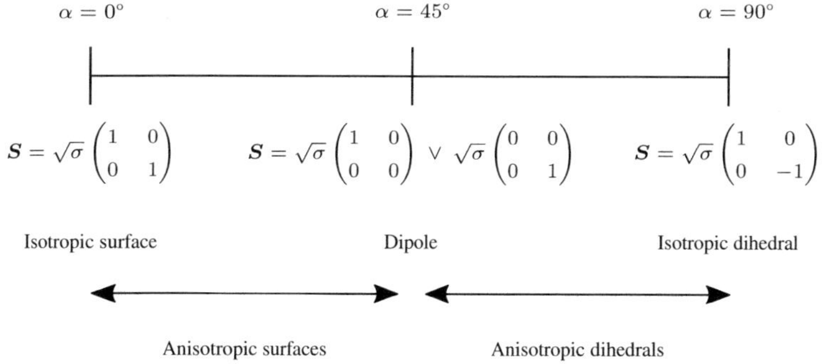

Figure 2.13: Schematic description of α parameter. Dipole scattering exemplary for horizontal orientation (left S-matrix) and for vertical orientation (right S-matrix) [Haj01].

seen that α is running from $0°$ to $90°$ and ranging from odd-bounce PEC surface scatterers (such as metallic spheres, plates, trihedrals) with an isotropic surface to dielectric surface scatterers with an anisotropic surface (i.e. where $S_{hh} \neq S_{vv}$) to dipole scattering (such as polarizing wire grids) and finally over dielectric dihedral scattering to perfectly conducting even-bounce scatterers (such as metallic dihedrals). It is to be noted that α also strongly depends on the angle of incidence (AOI) on the scatterer - for a fixed AOI α is increasing with increasing permittivity - and for a fixed permittivity α is increasing with increasing AOI [Clo09].
The angle β results to be half of the rotation angle θ of the scatterer around the radar LOS, as in Fig. 2.9 [Haj01]:

$$\theta = \beta/2 \ , \tag{2.31}$$

which means by the domain of the arctan function that a rotation of the target around the radar LOS can only be detected from $0°$ to $45°$. With this con-

straint, a measured dipole scatterer which is e.g. horizontally polarized will give the same value of the rotation angle θ as a vertically polarized dipole scatterer (namely 0°).
Finally, a schematic overview of a typical H/α map used in SAR polarimetry is presented in Fig. 2.14 [LP09]. The complete map shows that with increas-

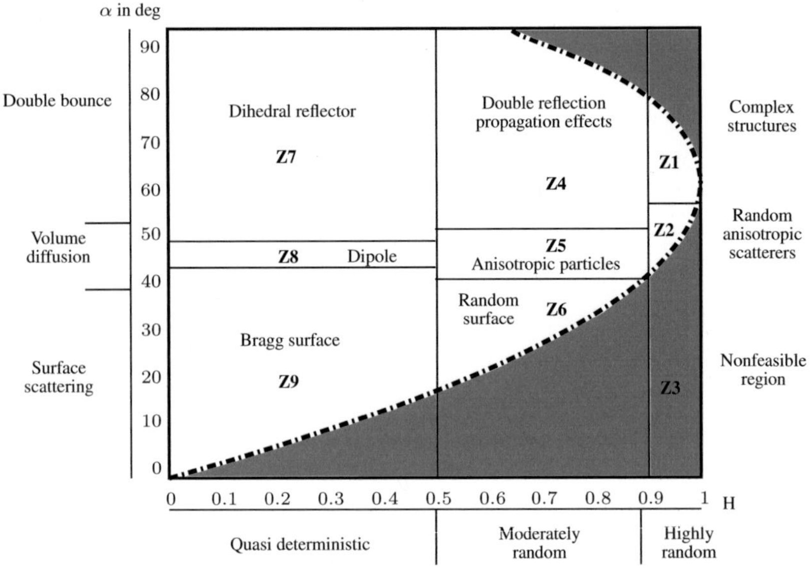

Figure 2.14: Schematic map of α-angle over entropy H with exemplary zones from 1-9 and descriptions, typically used in SAR polarimetry [LP09; Fen+15].

ing entropy, a deterministic polarimetric classification in terms of the α angle becomes more and more confined - up to the point of complete randomness, where the entropy equals one. At this point, only random anisotropic scattering can be detected. Furthermore, the map is divided into nine approximate zones, which are labeled in the figure. These zones can be understood as non-fixed and approximate classification regions that are especially useful in SAR polarimetry. For example, region Z9 describes general surface scattering which includes rough surfaces that can be described according to the Bragg or extended Bragg model, where increasing entropy goes along with increasing roughness [Clo09]. Even if the zones are just of approximate character, the connection to other fully polarimetric non-SAR radar systems is given, and the zones can still be taken as an approximate measure of target classification. Naturally, if the radar system or the application only provides measurement values that do not completely cover the entire H/α map, also other zones resp. zone border definitions can be defined according to the occurring measurement values, as can be seen e.g. in [AS13], where dielectric threat objects are classified. In this example, the measuring range of the utilized SAR MMW radar and the treated objects extends only up to a maximal entropy of 0.2 and a maximal α angle of $20°$.

It can be summarized that the entropy-alpha angle decomposition is helpful for classification of the polarimetric features in SAR data, yet it is not obvious if it is also feasible for non-SAR systems. For example, in automotive radar measurements the aspect angle changes rapidly contrary to earth-observing SAR measurements. The next chapters, especially Chapter 7, will indicate whether and where an implementation of the entropy-alpha angle decomposition is reasonable for automotive applications.

3 Experimental Fully Polarimetric MIMO FMCW Radar System at 77 GHz

In this chapter, the fully polarimetric experimental radar system used in this work is presented. A brief overview of the TDM operated MIMO FMCW radar design is given in form of a block diagram. Its dual polarized waveguide antenna with 8 Tx and 8 Rx elements is described in more detail. Simulation and measurement results of single antenna elements are compared. Parts of this chapter were published in [3].

3.1 Block Diagram

At first, the functionality of the complete radar system is described with the aid of a basic block diagram in Fig. 3.1. Inside the Tx module, a linear chirp generator creates fast frequency chirps over time. In this work, a chirp center frequency of 77 GHz is used. Each chirp signal is amplified by a power amplifier (PA) and switched consecutively over time to one of the n Tx antennas, i.e. one ramp is transmitted on each Tx channel. Once, Tx number n is reached, the whole process is repeated all over, starting again with Tx channel one. The Rx module is operated in a digital beam forming (DBF) way. This means, each of the m Rx channel signals is amplified by an own low noise amplifier (LNA) and then mixed down with the transmitted signal by an own mixer at that specific time. Also, each of the m received signals passes an own low-pass filter (to avoid aliasing and reduce the noise bandwidth) before it is sampled synchronously by an own analog-to-digital converter (ADC). All in all, m LNAs, low-pass filters and ADCs are used, one chain for each channel. Both modules are controlled by a field programmable gate array (FPGA) which is itself operated by a personal computer (PC). The FPGA is used to control the timing of the TDM switch, to set the modulation (chirp) parameters and also to set the sample rate/frequency.

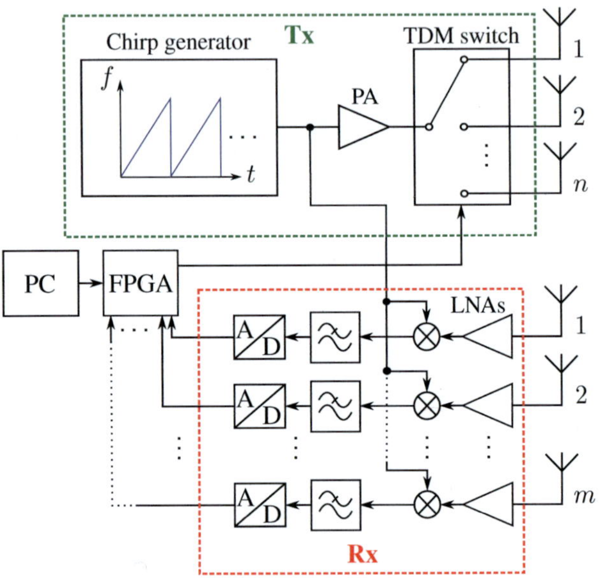

Figure 3.1: Coarse block diagram of complete radar sensor comprising a Tx and a Rx module.

3.2 8x8 Dual-Polarized Waveguide Antenna

According to Section 2.2, all that is needed to alter an existing single-polarized TDM MIMO radar into a fully polarimetric one is to change its antenna. An orthogonal polarized antenna enables the radar system to become fully polarimetric, as described in Section 2.2. For a high performance polarimetric radar it is essential that the polarization isolation between the single Rx antenna elements is high. The polarization purity is only determined by the Tx- and Rx-antennas of the radar [Bal05]. Therefore, the antennas must have a high cross-polar isolation. Due to the automotive field of application, the antenna is designed to have a larger field of view (FOV) in the azimuthal direction than in the elevation direction. One of the antenna types with the lowest polarization cross-talk that also has a broad beam pattern in one direction and a narrow beam pattern in the perpendicular direction is the sectoral horn antenna [Bal05]. Another advantage of this antenna type is its waveguide feed network with very low losses. The MIMO antenna presented here is a waveguide antenna with 16 sectoral horns. Half of the antenna elements are vertically-polarized E-plane sectoral horns and the other half are horizontally-polarized H-plane sectoral horns. A 3D-model of the antenna with indicated polarization directions and channel numbering is shown in Fig. 3.2.

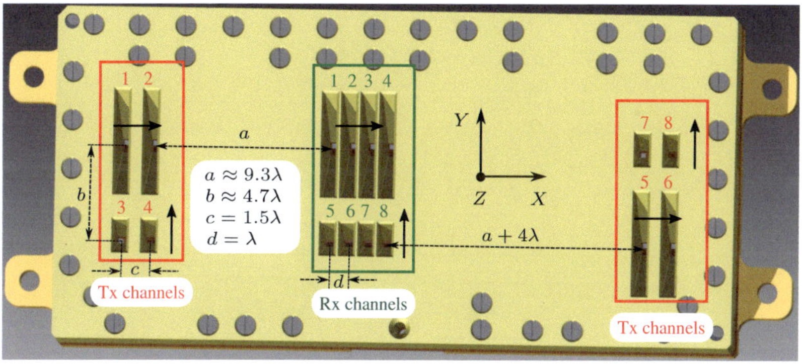

Figure 3.2: Rendered view of the 3D-model of the 8x8 dual-polarized MIMO waveguide antenna with aperture coordinate system and dimensions.

3.2.1 Virtual MIMO Array

Applying the result of Eq. (2.10) to the MIMO antenna dimensions from Fig. 3.2, the VA shown in Fig. 3.3 is created. It displays six distributed MIMO blocks which are separated in azimuth (x-direction) and elevation (y-direction). Each block is mainly a $\lambda/2$ ULA, only the elements at the edges/ends have a mutual spacing of one wavelength λ. In total there are 64 virtual channel combinations and every combination corresponds to a pair of Rx-Tx-antennas and their polarization combinations. The different colored symbols show the mnth polarization combinations and the numbers are consistent with the numeration in Fig. 3.2. It can be seen that the MIMO blocks on the x-axis have overlapping virtual positions. On the left, the horizontal-vertical (HV) and vertical-horizontal (VH) combinations overlap, whereas on the right, the HH and vertical-vertical (VV) combinations overlap. For these positions, any phase change of the received signal introduced by a certain DOA is equal. Therefore, on these overlapping positions, the polarimetric

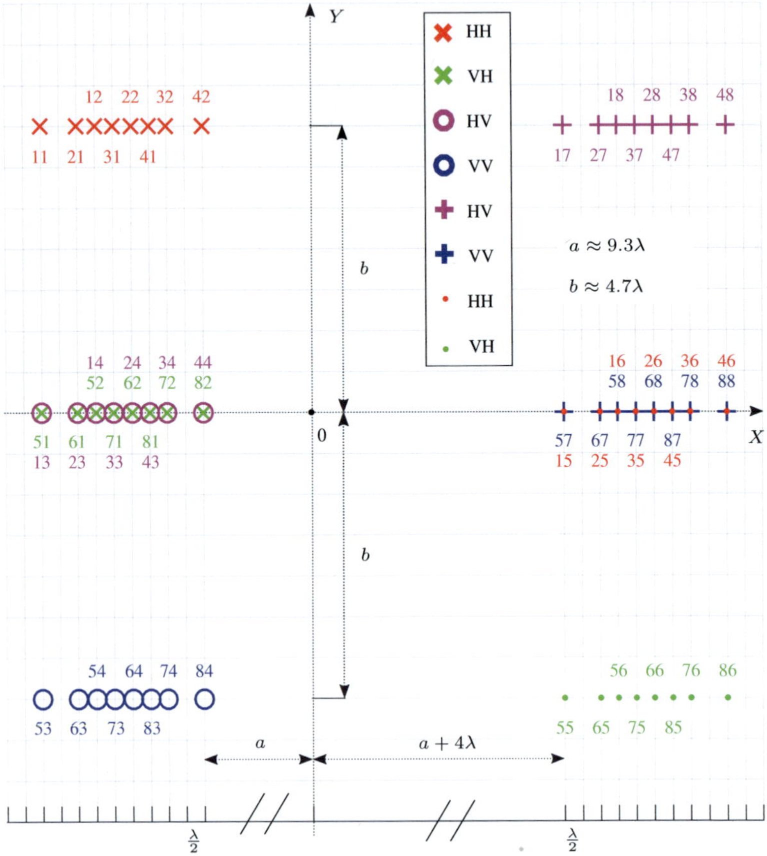

Figure 3.3: VA phase center positions of the MIMO antenna from Fig. 3.2 with polarization combinations (Rx-Tx).

phase differences between HV & VH and HH & VV resp. can be evaluated in any arbitrary scattering scenario.

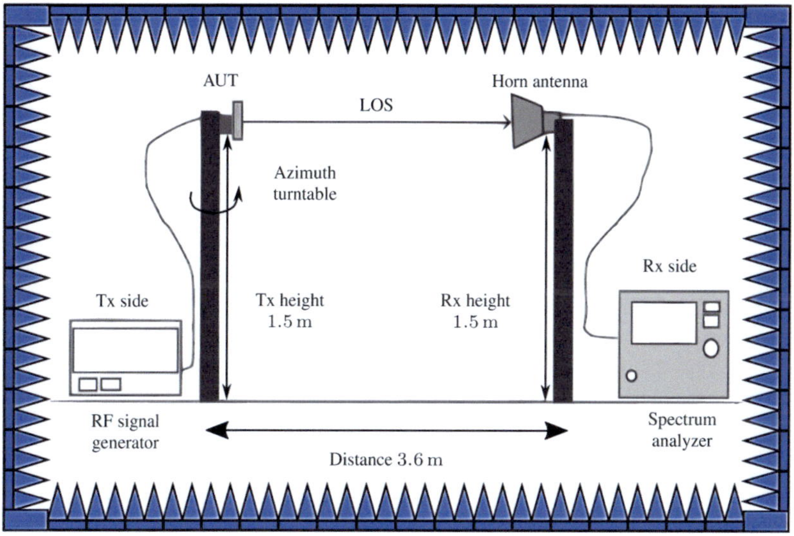

Figure 3.4: Schematic antenna measurement setup for the single antenna element characterization of the MIMO waveguide antenna shown in Fig. 3.2 [Ram+14].

3.2.2 Full-Wave Simulation and Antenna Pattern Measurements

In order to achieve equal radiation patterns in azimuth and elevation for both polarizations, the dimensions of both sectoral horn types were tuned in the 3D-full wave simulator CST MICROWAVE STUDIO® (CST) accordingly. The simulated normalized vertically- and horizontally-polarized azimuthal antenna patterns of the E-plane and H-plane horn antennas are plotted in Fig. 3.5. In azimuth, the vertically-polarized radiation patterns in Fig. 3.5a have a 3 dB-beamwidth of 60°. The simulation of the elevation patterns resulted in a smaller 3 dB-beamwidth of 30°. The cross-polar patterns show very low magnitudes and the cross-polar isolation is higher than 30 dB for almost the entire 3 dB-beamwidth. The simulation results show very similar pattern parameters also for the H-plane horn in Fig. 3.5b. The maximum gain amounts to 12.6 dBi for the vertically-polarized E-plane horn and to 12.7 dBi for the horizontally-polarized H-plane horn. With the parameters obtained by the simulation results, the waveguide antenna was fabricated using gold-plated aluminum as material. The single antenna elements were characterized in an anechoic chamber with the measurement setup introduced in Fig. 3.4. A radio frequency (RF) signal generator generates a continuous wave (CW) signal at 77 GHz which is transmitted by one of the connected sectoral horn antenna elements. The specific measured element is labeled as antenna under test (AUT). The entire waveguide antenna is mounted on top of an azimuth turntable. On the opposite side of the AUT in a LOS distance of approx. 3.6 m (which ensures the far field condition stated in [Bal05]), a standard pyramidal horn antenna, which is connected to a spectrum analyzer, is used as receiver. The gain of the standard horn amounts to 20 dBi. To obtain co-pol patterns, the pyramidal horn is brought into a co-pol position with the AUT and for x-pol measurements, the horn is rotated about 90° so that it is perpendicularly polarized to the AUT. By rotating the AUT in azimuthal direction, the antenna pattern can be measured. The received power for each rotation step is recorded. After the full pattern was measured, a second 20 dBi standard gain horn is mounted as AUT, so that a reference level is obtained with the spectrum analyzer. In comparison to this level, a maximal gain of approx. 12 dBi was measured for both sectoral horn type antenna elements which is in good agreement with the simulation results. The measured azimuthal antenna patterns are shown exemplary for one E-plane and one H-plane horn together with the corresponding simulation results in Fig. 3.5. The measurement results confirm the good agreement with the simulation results. The obtained co-pol patterns match almost perfectly with the simulation for both horns. Only the H-plane horn shows a slight ripple over its pattern which is most probably caused by the larger flare angle of the horn in comparison to the E-plane horn. The effect is augmented by fabrication tolerances and measurement setup inaccuracies like e.g. misalignments of the AUT and receiver horn antenna. In case of the E-plane horn, the E-field is tapered along the mouth of the horn. Thus, to reach a similar antenna gain, the flare angle can be reduced in comparison to the one of the H-plane horn, causing less ripple over the pattern [Bal05]. Comparing the x-pol patterns, larger deviations between simulation and measurement results can be seen. One reason is that very small inaccuracies have already a large impact at very small dB scales. Another reason is that fabrication tolerances and imperfections lead to slightly higher x-pol levels for both horn types as obtained from the ideal simulation environment. Also, mechanical misalignments of the mounted AUT and Rx horn can cause higher x-pol levels as would ideally be possible. Furthermore, the x-pol reception of the Rx horn is expected to be in the same order of magnitude as

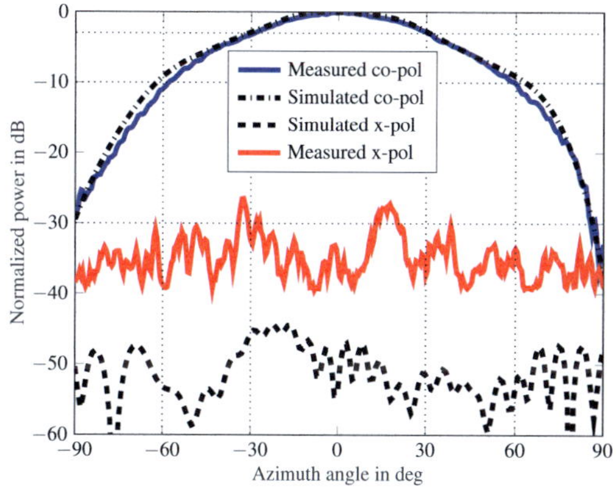

(a)

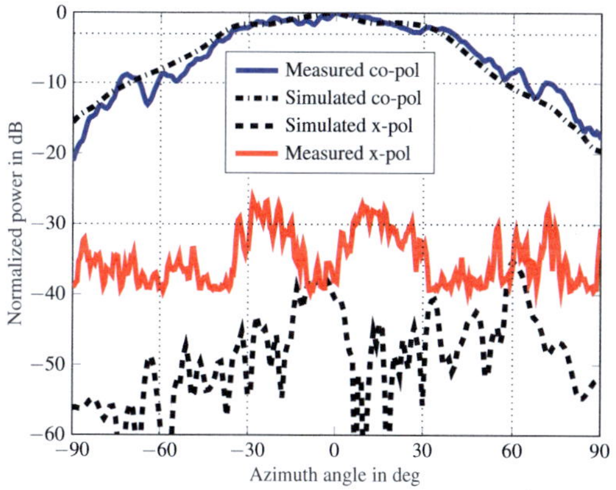

(b)

Figure 3.5: Comparison of measured and CST-simulated azimuthal normalized antenna patterns in dB. (a): Co-pol and x-pol patterns of E-plane horn (Tx 3). (b): Co-pol and x-pol patterns of H-plane horn (Tx 1).

the one of the AUT. Nevertheless, the measured x-pol levels are still very low for both horn types and the polarization isolation mostly exceeds 30 dB over the entire patterns.

3.3 Redesign of 8x8 Dual-Polarized Waveguide Antenna

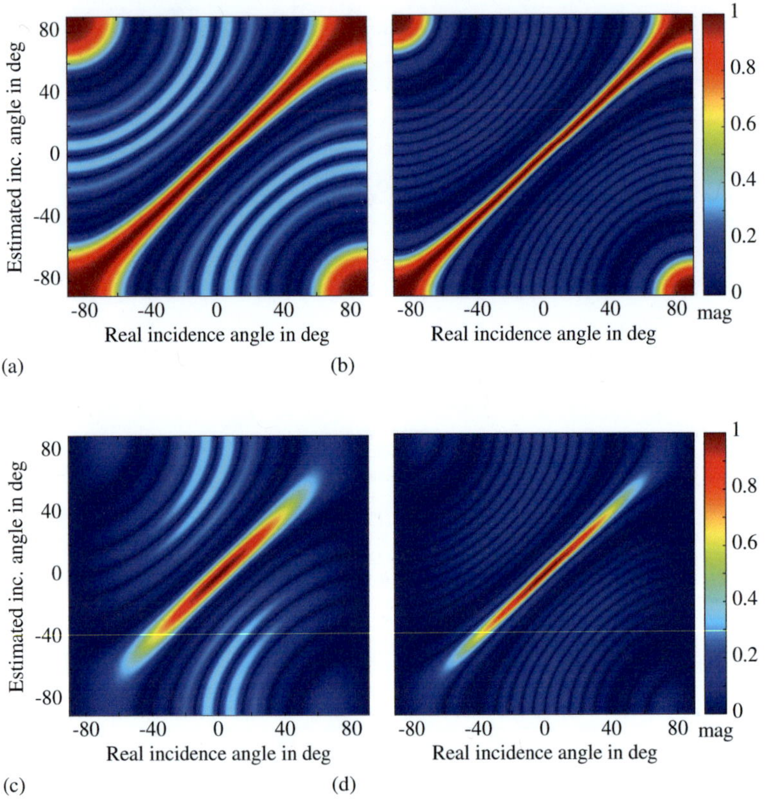

Figure 3.6: Ambiguity functions of one 8-element quasi-ULA from the antenna in Fig. 3.2 with $q = 0$ in (a) and $q = 1.8$ in (c), and of one 16-element quasi-ULA from Fig. 3.7 with $q = 0$ in (b) and $q = 1.8$ in (d).

One quasi-ULA MIMO block with 8 elements of the VA in Fig. 3.3 is considered, and the DOA estimated angle is calculated for all possible occurring angles between $-90°$ and $90°$ by the method described in Section 2.3.2. Holes in the array are filled with zeros. In order to investigate the beamforming performance of the array with respect to grating and side lobes, an angle ambiguity function can be defined by plotting the real input AOA over the estimated AOA (by taking the maximum of the angle spectrum in Eq. (2.13)). In the following, the simple $\cos(\varphi)^q$ pattern approximation function from [HE07] has been chosen to model the single array element's antenna patterns in azimuthal direction. With increasing q factor, this model generates a pattern function with increasing directivity. In Fig. 3.6a and Fig. 3.6c, the angle ambiguity functions are presented for a q factor of 0 (isotropic antenna elements) and 1.8 (approx. matching beam patterns with the co-pol sectoral horn patterns shown in Fig. 3.5 from $-3\,\mathrm{dB}$ to $3\,\mathrm{dB}$). As can be seen in both plots, a conspicu-

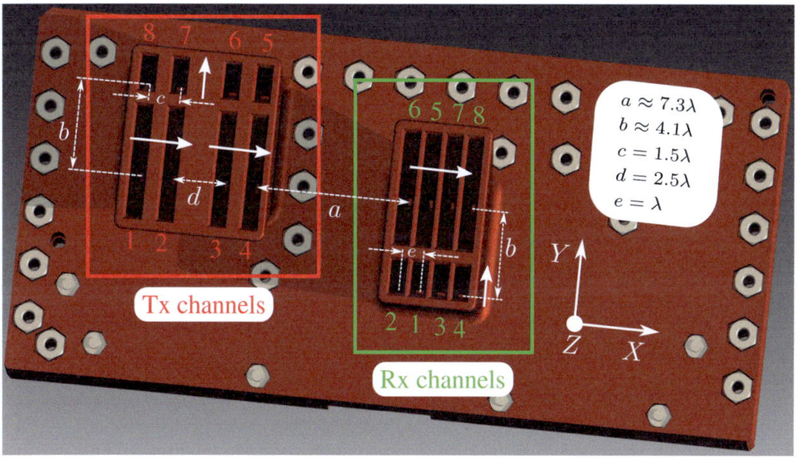

Figure 3.7: Rendered view of the 3D-model of the redesigned 8x8 dual-polarized MIMO waveguide antenna with aperture coordinate system and dimensions.

ous side lobe level exists, most probably caused by the two gaps at the edges of the quasi-ULA. Furthermore, the main lobe is quite broad resulting in a low angle resolution. In total, two gaps in the array preponderate in relation to only 8 elements in total. By doubling the size of the quasi-ULA aperture to a total of 16 elements (while keeping the two gaps at the edges) the resolution is improved and the side lobe levels are decreased significantly, as is demonstrated in Fig. 3.6b and Fig. 3.6d for a q factor of 0 and 1.8, respec-

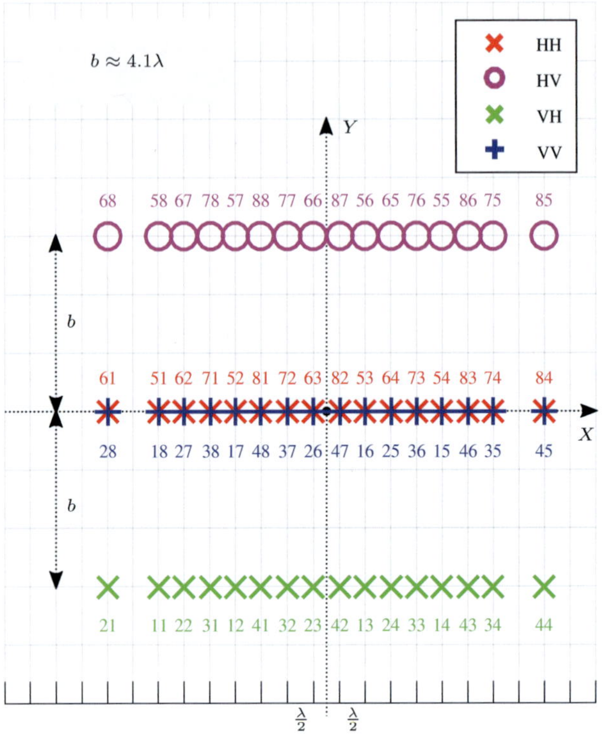

Figure 3.8: VA phase center positions of the MIMO antenna from Fig. 3.7 with polarization combinations (Rx-Tx).

tively. Following this intention, a redesign of the dual-polarized waveguide antenna from Fig. 3.2 is constructed. A 3D model of the redesigned antenna is shown in Fig. 3.7. The total number of elements is maintained in the second version, just a rearrangement of the Rx and Tx blocks was carried out, keeping also the same sectoral horn antenna elements. Calculating the corresponding VA by using Eq. (2.6), its Rx-Tx antenna combinations are presented with polarization directions and numbering in Fig. 3.8. In this redesigned VA, only three MIMO blocks are created, separated in elevation direction. The upper block comprises HV-polarization combinations whereas the lower block comprises VH-polarization combinations. Due to reciprocity, the information obtained with these two blocks is very similar, and almost the same in reasonable distances, where the elevation separation can be neglected. Also, the phase difference between the HV- and VH-polarization combinations plays a minor part. Focus for the redesign is thus specifically laid on overlapping HH- and VV-polarization combinations as the phase differences are important to distinguish between odd and even bounce scattering.

4 Polarimetric Calibration Methods

One difference between the virtual array (VA) and the real physical MIMO antenna consists of the design related path length variations of the Rx-antenna connections to the RF-board. Also, small deviations in the monolithic microwave integrated circuits (MMICs) and other differences such as manufacturing tolerances etc. on the RF-board make it impossible to obtain a coherent reception of the incoming signals at higher frequencies. It is therefore inevitable to perform a calibration of the radar system. In literature, mainly three different approaches are discussed, the general calibration technique (GCT) [Whi+91], the single-reference general calibration technique (SRGCT) [WK91] and the isolated antenna calibration technique (IACT) [SUT90]. These approaches differ mainly in their levels of complexity and the number of needed calibration targets. In this chapter, calibration methods are discussed and a single-reference calibration is performed using measurements in an anechoic chamber. All measurements presented in this chapter were performed with the fully polarimetric radar sensor from Chapter 3 with the antenna from Section 3.2 and the FMCW modulation parameters listed in Appendix A.2. Parts of this chapter were published in [3].

4.1 Calibration Based on the IACT

One of the most basic and intuitive but yet accurate polarimetric calibration approach is the IACT. In this approach, it is assumed that the used antennas have a high polarization isolation. This condition is fulfilled for the sectoral horn antennas in this work, cf. Section 3.2.2. In order to obtain the real scattering matrix from the measured scattering matrix $\boldsymbol{S_m}$, two correction matrices for Rx & Tx are formulated, $\boldsymbol{R}$ and $\boldsymbol{T}$:

$$\boldsymbol{R} = \begin{pmatrix} R_h & 0 \\ 0 & R_v \end{pmatrix} \qquad \boldsymbol{T} = \begin{pmatrix} T_h & 0 \\ 0 & T_v \end{pmatrix} . \tag{4.1}$$

The entries on the off-diagonals describe the coupling between the polarizations and are set to zero. Thus, there are only 4 correction terms instead of 8 needed for the GCT and the SRGCT. Nevertheless, the IACT reaches almost the same quality for systems with a high polarization isolation as the much more complex GCT [Whi+91]. A background subtraction [WK91] is omitted in this work because the background changes in each measurement and thus the method is prone to errors. With these corrections, the real S-matrix can be approximated:

$$\boldsymbol{S} = \boldsymbol{R} \cdot \boldsymbol{S_m} \cdot \boldsymbol{T} = \begin{pmatrix} R_h S_{hh} T_h & R_h S_{hv} T_v \\ R_v S_{vh} T_h & R_v S_{vv} T_v \end{pmatrix} . \tag{4.2}$$

In [SUT90], the considerations are based on a monostatic system. In order to generalize the IACT for quasi-monostatic and bistatic MIMO systems, the calibration formulas from [SUT90] have to be adapted. Given the received scattered power from a point target from the *n*th Tx- at the *m*th Rx-antenna $P^r_{mn} = (\hat{U}^r_{mn})^2/2Z_a$, where $\hat{U}^r_{mn}$ is the *mn*th peak voltage and Z_a is the *n*th Rx-antenna impedance. This is inserted into the bistatic two-way radar equation from [Avi13]:

$$\hat{U}^r_{mn} = \sqrt{\frac{P^t_n G^t_n G^r_m \lambda^2 2 Z_a}{(4\pi)^2 \, r^2_{txt} \, r^2_{trx}} \frac{\sigma_{mn}}{4\pi}} = \xi \, \frac{1}{r_{txt} r_{trx}} \sqrt{\frac{\sigma_{mn}}{4\pi}} \, . \tag{4.3}$$

In this equation, P^t_n is the power from the *n*th Tx, G^t_n and G^r_m are the *n*th Tx- and *m*th Rx- antenna gains, resp. and σ_{mn} is the corresponding *mn*th RCS. With the relation between scattering matrix and RCS [LP09; Mot06]:

$$\sqrt{\frac{\sigma_{mn}}{4\pi}} \frac{1}{r_{trx}} = \frac{S_{mn}}{\sqrt{4\pi} r_{trx}} e^{-j\vec{k}_i \cdot [2\vec{r}_t - (\vec{r}_{tx} + \vec{r}_{rx})]} , \tag{4.4}$$

the received peak voltage at the *m*th Rx can be formulated as:

$$\hat{U}^r_{mn} = \xi \, \frac{S_{mn}}{\sqrt{4\pi} r_{trx}} \frac{1}{r_{txt}} e^{-j\vec{k}_i \cdot [2\vec{r}_t - (\vec{r}_{tx} + \vec{r}_{rx})]} . \tag{4.5}$$

Considering Fig. 2.4 in the far-field, where $r_t \gg (r_{tx}, r_{rx})$ (quasi-monostatic case), the following simplifications can be made:

$$r_{txt} \approx r_{trx} \approx r_t = r \quad \text{and} \quad \vec{k}_i \parallel k_0 \vec{e}_z \,, \tag{4.6}$$

where $\vec{e}_z$ is the unit vector in z direction. The received calibrated peak voltage can then be written as:

$$\hat{U}^r_{mn} = \xi \, \frac{1}{\sqrt{4\pi r^2}} R_m S_{mn} T_n e^{-2jk_0 \cdot r} \,. \tag{4.7}$$

To calibrate all entries of the measured S-matrix, one co-pol target and one x-pol target have to be measured. Furthermore, only the RCS of the co-pol target needs to be known. The simplest co-pol target with a theoretically well-known RCS is the sphere. Thanks to its geometry, it is also a target which is insensitive to alignment errors. If the diameter d_0 of the sphere is much larger than the wavelength, the RCS of the sphere is $\sigma_0 = \pi(d_0/2)^2$ [Avi13] and its scattering matrix is (cf. Eq. (2.14a)) [Mot06]:

$$\boldsymbol{S_0} = \sqrt{\sigma_0} \begin{pmatrix} 1 & 0 \\ 0 & 1 \end{pmatrix} \,. \tag{4.8}$$

For the x-pol target, rotated dipoles, cross-polarizers or rotated dihedrals come into consideration [SUT90]. These targets have in common that their alignment is difficult, not least because they have to be rotated by a certain angle around one of their axes while the other 2 axes need to remain fixed. However, in this work a $45°$ rotated dihedral which is carefully aligned, is used for the calibration procedure. The corresponding S-matrix $\boldsymbol{S_d}$ of a dihedral with its seam rotated by an angle θ is given by Eq. (2.15). It can easily be seen that at a $\theta = 45°$ position, the dihedral represents a pure x-pol target. After measuring the sphere and the rotated dihedral at a boresight position, the uncoupled correction terms R_m and T_n can be extracted from the measured S-matrices using Eq. (4.7). Two physical reasonable assumptions are then made. First, the antenna parameters (contained in ξ from Eq. (4.3)) are assumed to be equal for all Tx- and Rx-antennas. Second, the x-pol target is assumed to be reciprocal, thus $S_{hv} = S_{vh}$ (strictly valid only for monostatic systems [LP09]).

Measuring an unknown target, its calibrated scattering matrix $\boldsymbol{S_u}$ can then be determined by using the following equations:

$$S^u_{vv} = \frac{\hat{U}^u_{vv}}{\hat{U}^0_{vv}} \left(\frac{r^u_{vv}}{r^0_{vv}}\right)^2 S_0 \; e^{2jk_0\left(|r^u_{vv} - r^0_{vv}|\right)} \tag{4.9a}$$

$$S^u_{hh} = \frac{\hat{U}^u_{hh}}{\hat{U}^0_{hh}} \left(\frac{r^u_{hh}}{r^0_{hh}}\right)^2 S_0 \; e^{2jk_0\left(|r^u_{hh} - r^0_{hh}|\right)} \tag{4.9b}$$

$$S^u_{hv} = \hat{U}^u_{hv} \frac{1}{\sqrt{\hat{U}^0_{vv}\hat{U}^0_{hh}}} \frac{{r^u_{hv}}^2}{r^0_{vv} r^0_{hh}} K S_0 e^{2jk_0 r^u_{hv}} \sqrt{\frac{\beta}{\alpha}} \tag{4.9c}$$

$$S^u_{vh} = \hat{U}^u_{vh} \frac{1}{\sqrt{\hat{U}^0_{vv}\hat{U}^0_{hh}}} \frac{{r^u_{vh}}^2}{r^0_{vv} r^0_{hh}} \frac{1}{K} S_0 e^{2jk_0 r^u_{vh}} \sqrt{\beta\alpha}\,, \tag{4.9d}$$

with $K = \left(r^d_{vh}/r^d_{hv}\right) \cdot \left(\hat{U}^d_{vh}/\hat{U}^d_{hv}\right)^{1/2}$, $\alpha = e^{-2jk_0\left(|r^d_{vh} - r^d_{hv}|\right)}$ and $\beta = e^{-2jk_0(r^0_{vv} + r^0_{hh})}$. The superscripts are used as follows: u for unknown target, d for dihedral (calibration) target, 0 for (calibration) sphere target. The calibration formulas only depend on the measured voltages, the distances of calibration and unknown targets from the radar, and the theoretically calculated value of S_0. Given the calibration formulas in Eqs. (4.9a) to (4.9d), the calibration can be executed. Since the radar system is a FMCW radar, the formulas are applied in the range-velocity spectrum, i.e. after applying FFTs to the time signals (cf. Appendix A.1). Thus, the amplitude, phase and range bin of the target peak value in the range-velocity spectrum are stored. This process is done with the sphere and the 45° rotated dihedral. To measure unknown targets, the same procedure is followed and the calibration formulas are applied to determine the calibrated S-matrix. The calibration measurements are performed in an anechoic chamber, the setup is shown in Fig. 4.1. The calibration sphere has a diameter of $d_0 = 30\,\text{cm}$ and is positioned at boresight in $3.9\,\text{m}$ distance from the radar. The calibration dihedral has a width of $w = 18.5\,\text{cm}$, a height of $h = 20\,\text{cm}$ and is also placed at boresight in $2.2\,\text{m}$ distance.

The need for calibration can be easily understood if Fig. 4.2 is considered. The plot shows the uncalibrated phase values for all channels of the measured

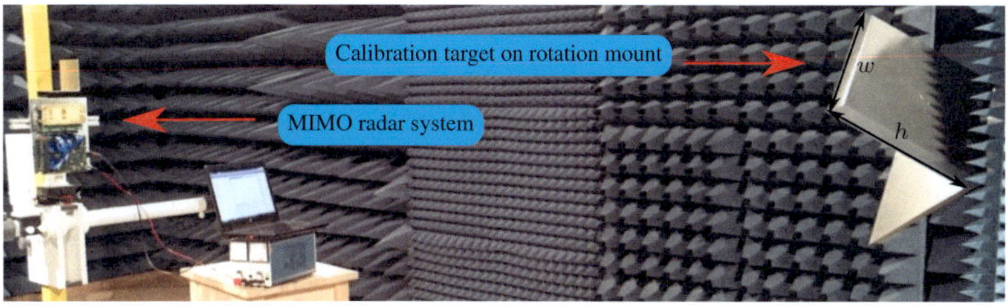

Figure 4.1: Calibration measurement setup with a 45° rotated dihedral as x-pol calibration target in an anechoic chamber.

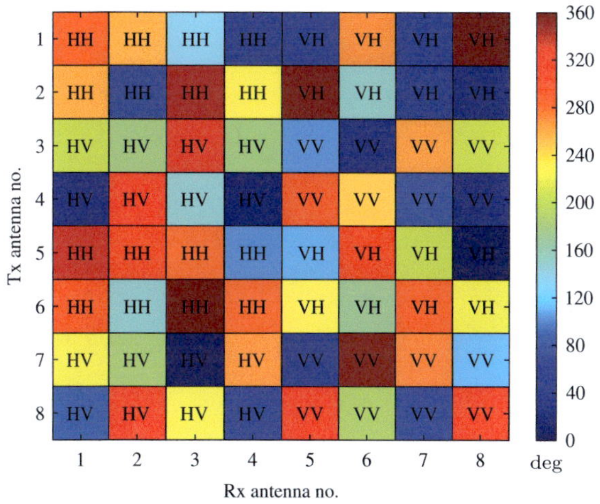

Figure 4.2: Uncalibrated phase values of calibration target measurements over all virtual channels for the calibration sphere.

calibration sphere. It can be seen that the phase values for all co-pol as well as x-pol channels are random. Thus, no reasonable DOA estimation and/or evaluation of polarimetric phase features can be conducted.
In order to evaluate the quality of the calibration, the calibration targets are considered as unknown targets and are calibrated with their own measurement data. This means that the calibration data is used on itself to assess on the one hand the calibration correctness of the magnitudes and phases and on the other

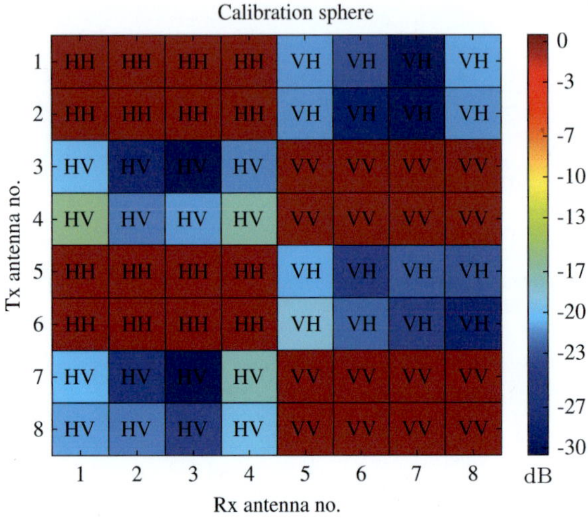

(a)

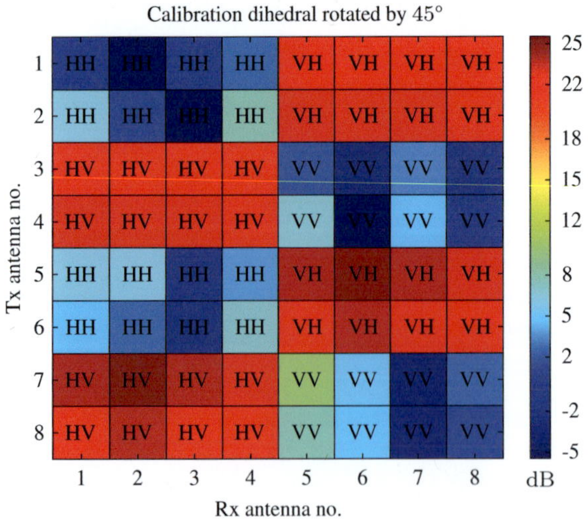

(b)

Figure 4.3: Calibrated S-matrix magnitude values of calibration target measurements over all virtual channels (normalized to the theoretical value S_0 of the calibration sphere). (a): Cal. sphere. (b): Cal. dihedral rotated by 45°.

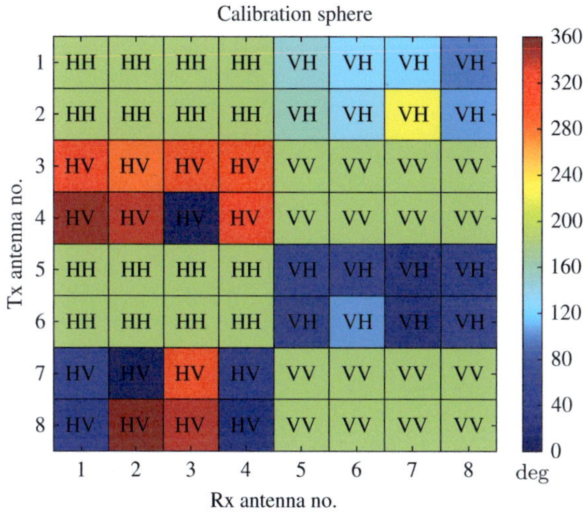

(a)

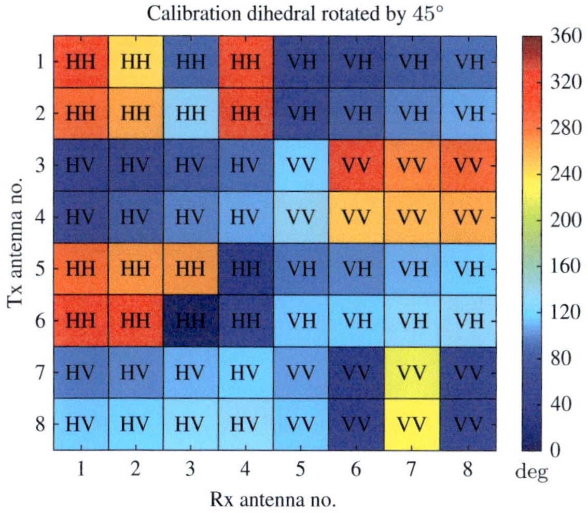

(b)

Figure 4.4: Calibrated phase values of calibration target measurements over all virtual channels. (a): Cal. sphere. (b): Cal. dihedral rotated by $45°$.

hand the single-reference performance on the x-pol channels, since the RCS of the x-pol calibration target was not used during the calibration procedure. A plot of the calibrated magnitudes of all virtual channels from the calibration sphere and dihedral is shown in Fig. 4.3. The magnitudes are normalized to the theoretical RCS magnitude of the sphere. It can be observed that the magnitudes of the calibrated sphere target are equal on all co-pol channels with the normalized value of 0 dB. The lowest x-pol suppression in channel (1,4) amounts to 16 dB. The averaged magnitude ratio of all co-pol to all x-pol channels is 23 dB. The normalized magnitudes of the $45°$ rotated dihedral are also almost equal with 23.5 dB on all x-pol channels, with a slight deviation of ± 1.5 dB on some channels. In this case, the lowest co-pol suppression occurs in channel (5,7) with a value of 12 dB. The averaged magnitude ratio of all co-pol to all x-pol channels is -20.3 dB. In a similar way, the calibrated phase values are plotted in Fig. 4.4. The phases on all co-pol channels for the calibration sphere are well calibrated and show an equal absolute value of $180°$. Thus, any phase differences (relative phases) between overlapping HH and VV channels are zero, as expected. In case of the dihedral calibration target, the phases show a small progressive phase shift towards rising Rx-channel numbers. These deviations probably relate to small positioning errors of the dihedral - it had not been aligned exactly in the same azimuth position as the sphere with which it was calibrated (Eqs. (4.9c) and (4.9d)). Nevertheless, the phase differences between overlapping HV and VH channel combinations are also zero as expected.

4.2 General Calibration Techniques

As stated in the former section, general calibration techniques can be used to calibrate fully polarimetric radar systems with the best performance, even if the antennas do not have a high x-pol isolation. Basically, two of these methods, the mentioned GCT [Whi+91], and the SRGCT [WK91] can be compared. Both methods aim for completely filled correction matrices $\boldsymbol{R}$ and $\boldsymbol{T}$ from Eq. (4.1), with 8 correction terms instead of only 4, for Rx and Tx respectively, thus being able to correct also the polarization cross-talk between the antennas, in contrast to the IACT. While for the SRGCT also only one reference target with a theoretically known RCS is needed, for the GCT three targets with calculated RCS must be used. For both methods, an optional background subtraction can be performed, as for the IACT. The three different targets needed include a co-pol target (most convenient is a sphere), a strongly cross-polarizing target (most convenient is a $45°$ rotated dihedral) and additionally a strongly polarizing target such as a polarizing wire grid or alternatively a dihedral rotated by $0°$ or $90°$. Important is that the scattering matrices of these three canonical objects are linearly independent (cf. Eq. (2.14)). The salient feature of the SRGCT is that it uses e.g. the co-pol target as a reference to calibrate the RCS of the other two targets, similar to the IACT. Because of the double number of correction terms, both general calibration techniques are much more complex than the IACT and are based on matrix inversions and eigenvalue decompositions.

All in all it can be stated that both general calibration techniques yield the best results concerning accuracy and x-pol suppression (in the order of $50\,\mathrm{dB}$ or higher). However, for antennas with inherently high x-pol suppression like the sectoral horn antennas used in this work, almost identical results are obtained, only that the total x-pol suppression in case of the IACT yields weaker results (in the order of $35\,\mathrm{dB}$ or higher), due to neglection of the polarization coupling terms in the correction matrices [Whi+91]. Because of the minor expected improvements over the IACT in case of antennas with high x-pol isolation and the much higher complexity, the general calibration techniques were not tested and are thus not further discussed in this work.

5 Fixed Sensor Multi-Angle Measurements in Anechoic Chamber

In this chapter, the calibration method obtained in Section 4.1 is verified and tested extensively based on exemplary measurement results of different canonical objects in an anechoic chamber. Pauli-decomposed results are shown in a color-coded format and a statistical analysis of the measurement results of different canonical objects in varying positions is conducted in order to show the performance of the calibration method. On the basis of these measurement results, the feasibility of the IACT for practical use is demonstrated. Additionally, rotational measurement results of selected two-wheeled vehicles are demonstrated in the Pauli-decomposed and color-coded format to demonstrate the added value of the polarimetric information. Finally, an analysis of polarimetric multipath and DOA detection is conducted, using canonical object measurements in an anechoic chamber in order to test the potential of multipath mitigation using polarimetry. All measurements presented in this chapter were performed in an anechoic chamber using the fully polarimetric radar sensor from Chapter 3 with the antenna from Section 3.2 and the FMCW modulation parameters listed in Appendix A.2. Parts of this chapter were published in [3; 4; 2; 7].

5.1 Statistical Evaluation of Arbitrarily Located Canonical Objects

In order to evaluate the quality of relative polarimetric phase extraction (cf. Eq. (2.4)) and co-pol to x-pol magnitude ratios after the calibration, different canonical x-pol and co-pol even and odd bounce objects were measured. In total, 97 different measurements from a list of canonical objects, stated in Table 5.1, were performed and a statistical evaluation was carried out.

Object(s)	Pol. type	Rot. angle θ	Dim(s) in cm
Dihedrals	co-pol even bounce	$0°$, $22.5°$, $67.5°$, $90°$	(w, h): $(20, 20)$, $(18.5, 20)$, $(10, 14)$
Square rough surface plate	co-pol odd bounce	—	edge lengths: 28
Sphere	co-pol odd bounce	—	diameter: 30
Trihedrals	co-pol odd bounce	—	edge lengths: 10, 12.5, 18, 35
Dihedrals	x-pol	$22.5°$, $45°$, $67.5°$	(w, h): $(20, 20)$, $(18.5, 20)$, $(10, 14)$

Table 5.1: List of measured canonical objects with polarimetric type, dimensions and rotation angles according to Fig. 2.9. The objects were randomly placed in distances from 2 m-6 m and at varying azimuth and elevation directions of up to $\pm$ 25°.

Even bounce objects ideally exhibit a 180° phase-shift between their HH and VV S-matrix elements. In contrast to that, odd bounce objects theoretically exhibit no phase-shift between their HH and VV S-matrix elements. Finally, also no phase shift occurs between the HV and VH S-matrix components for x-pol objects in the (quasi-)monostatic case due to reciprocity of the S-matrix [LP09]. Pure mixed objects comprise a co-pol even bounce part to the same extent as a x-pol part (cf. Eq. (2.16)). All objects were placed in distances ranging from 2 m-6 m and at boresight direction or at off-broadside azimuth and elevation directions of up to $\pm$ 25°. A recapitulation on the scattering behavior of canonical objects is given in Section 2.5.

Subsequently, the co-pol phase differences $\Delta\phi$ between HH and VV for the even and odd bounce targets and between HV and VH for the x-pol targets of all concerning overlapping MIMO-channels on the x-axis of the VA in Fig. 3.3 were averaged. In Table 5.2, the statistical evaluation of the measurement results is presented. As additional information to the Table 5.2, the average co-pol to x-pol ratios remained well above/below ± 10 dB for all pure co- or x-pol measurements with a mean value of 21.6 dB for the co-pol targets

Objects and phase-shift $\Delta\phi$	Cross-polar $\lvert\phi_{HV} - \phi_{VH}\rvert$	Odd bounce $\lvert\phi_{VV} - \phi_{HH}\rvert$	Even bounce $\lvert\phi_{VV} - \phi_{HH}\rvert$
Mean of $\Delta\phi$	$16.25°$	$13.27°$	$166.91°$
Std. dev. of $\Delta\phi$	$8.12°$	$6.81°$	$7.86°$
Max. of $\Delta\phi$	$37.4°$	$27.55°$	$177.91°$
Min. of $\Delta\phi$	$5.22°$	$2.83°$	$148.5°$
No. of meas.	33	26	38

Table 5.2: Statistical evaluation of calibrated phase-shift measurements of different canonical odd bounce, even bounce and cross-polar objects from the list in Table 5.1.

and $-19\,\text{dB}$ for the x-pol targets. Thus, the canonical objects can clearly be classified into co- and x-pol targets by the ratio of co- to x-pol magnitudes. Furthermore, the even and odd bounce scattering processes can distinctly be detected from the relative phase-shifts. The standard deviations (SDs) of the average relative phase-shifts between HH and VV show that, although there is a certain spread in the values, the bounce processes are reliably detectable in any case. The mean value and the SD of the phase-shift between VH and HV of all measurements is also approaching the concerning values of the odd bounce measurements. The reason for deviations in the phase-shifts are varying RCS values of the measurements, probably caused by misalignment and the non-ideality of the objects, as already stated in Section 4.1 for the x-pol calibration target.

5.2 Rotational Measurements of Canonical Objects

After the feasibility of the IACT was demonstrated by using a statistical evaluation approach, selected rotational measurement results of canonical objects in an anechoic chamber are shown. As the most versatile canonical object, a dihedral was rotated about its seam and measured at each angle rotation step.

Also, vertical and horizontal strip-grid polarizers were rotated in azimuthal direction and measured in the same way. In Fig. 5.1, the measurement setup for the self-rotational measurements of the dihedral scatterer is shown.

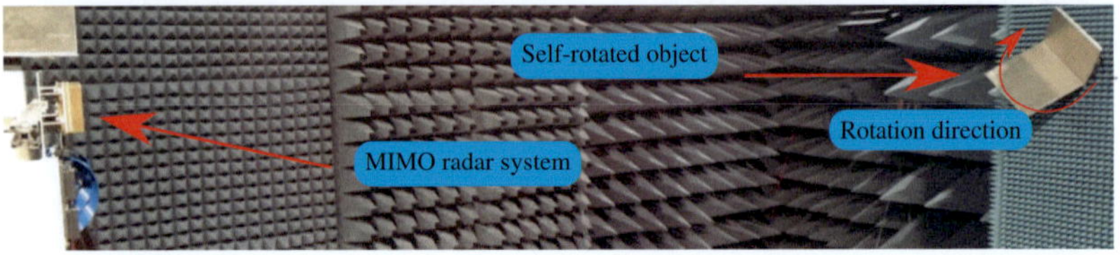

Figure 5.1: Measurement setup with dihedral object rotated around its seam (rotated in elevation direction).

Subsequent to applying the Pauli decomposition, which is introduced in Section 2.6.1, to the calibrated measurement data, the RCS of the different canonical scattering processes is visualized in a color-coded red-green-blue (RGB) format. The first three Pauli components are assigned to the RGB colors, where the RCS of component a is coded as blue, the RCS of component b as red and the cross-polar RCS $|c|^2/2$ is coded as green. By this method, an RGB image is created. In order to avoid noise in the images, the Pauli parameters obtained from each overlapping MIMO channel of the central ULAs in Fig. 3.3 are averaged, normalized to the overall maximum, and a threshold of $-40\,\mathrm{dB}$ is introduced. Below this value, all RCS values are set to the threshold which is illustrated as black color (RGB value $[0,0,0]$). If all three canonical scattering processes are equally present in a pixel, this results in white color (RGB value $[1,1,1]$) in the image. The brightness of each color illustrates the strength of RCS of the corresponding Pauli component ranging from 0 to 1. An example of the described additive RGB color scheme is shown in Fig. 5.2. For example, a polarizer yields magenta color in this format, since it is a mix of blue and red color. This can easily be seen, if the S-matrix in Eq. (2.14d) and the Pauli parameters in Eq. (2.19) are considered. In Fig. 5.3, the color-coded results of a self-rotated dihedral, located in boresight direction at a fixed distance of approx. $5.3\,\mathrm{m}$ are visualized. If compared to Fig. 2.9, the scattering behavior of the dihedral over its rotation angle θ can be very well followed. At $0°$, the seam of the dihedral is horizontally aligned, resulting in pure even bounce scattering, represented as red color. Due to the high range resolution of $7.5\,\mathrm{cm}$ (according to the measurement bandwidth of $2\,\mathrm{GHz}$ [Hua+11]) and

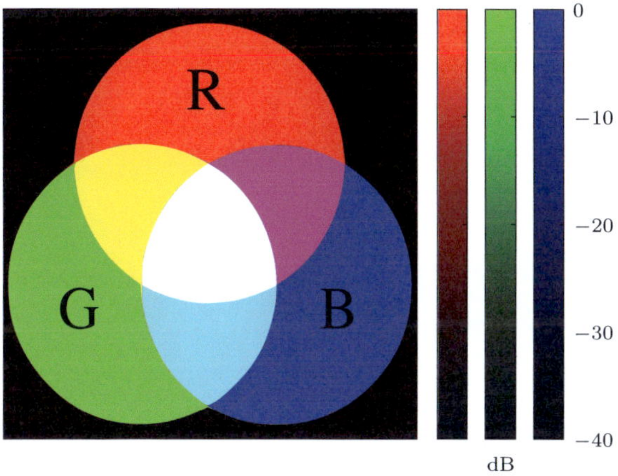

Figure 5.2: Additive RGB color scheme with logarithmic scale in dB.

the large size of the dihedral ($w = 18.5\,\mathrm{cm}$, $h = 20\,\mathrm{cm}$), a small part of odd bounce scattering in blue color can be observed, originating from the edges and seam. Also, a very small part of polarizing behavior (magenta color) can be observed, originating also from the edges and seam of the dihedral. Rotating it further, yellow color (mix of red and green) appears, which represents a mixed target (cf. Eq. (2.16)), having its maximum at $22.5°$, followed by green, which represents a pure x-pol target, having its maximum at $45°$ rotation angle (cf. Eq. (2.14c)). Eventually, at $90°$ rotation angle, the dihedral exhibits its maximum of co-pol even bounce scattering again, represented by red color. Then, the whole process is mirrored and the color gradient can be repeatedly observed in the opposite direction, up to a rotation angle of $180°$.

These results clearly demonstrate the high resolution in range direction and the fine color gradient (as a result of the high Pauli parameter separation capability) achievable with the presented radar system and the applied calibration, yielding the expected results obtained from the theory of the self-rotated dihedral in Eq. (2.15). It can be concluded that a physical dihedral represents an almost ideal canonical object, at least in terms of the scattering behavior between its edges and seam. Thus, for a radar system with low range-resolution, a dihedral with large height compared to one range bin is favorable, in order to minimize the mix of edge and seam scattering with the even bounce scat-

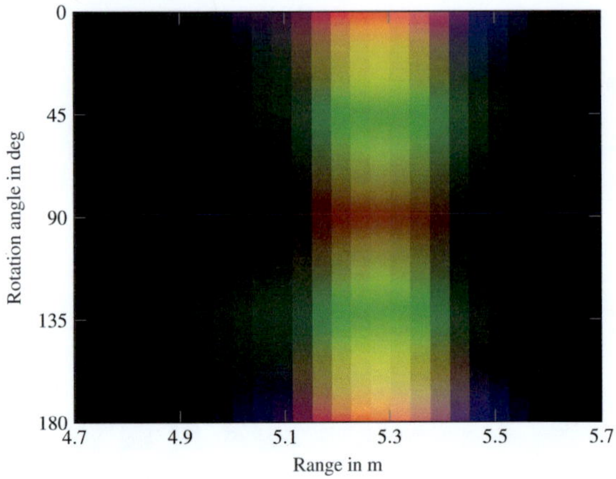

Figure 5.3: Pauli decomposed and RGB color-coded results of a $w = 18.5 \times h = 20\,\mathrm{cm}$ dihedral rotated about its seam (cf. Fig. 2.9).

tering in between them. Furthermore, a sharpening of the edges reduces their backscattering [WK91].

Next, the ability of the radar concerning detection of polarizer orientation was investigated. For the following results, a planar strip-grid polarizer was designed and simulated in CST, and then manufactured and measured. Gold-plated copper strips with a width of $300\,\mu\mathrm{m}$ and a spacing of $500\,\mu\mathrm{m}$ yielded very good simulation results. The strips were etched on a $50\,\mu\mathrm{m}$ thin $24\,\mathrm{cm} \times 24\,\mathrm{cm}$ Kapton foil. Because the dimensions of the strips, their spacing and the foil thickness are much smaller than the wavelength at $77\,\mathrm{GHz}$ $(3.9\,\mathrm{mm})$, the influence of the foil can be neglected. Ideally, for a very fine strip grid, backscattering only occurs in the dimension parallel to the strips [Gol98]. The foil was stretched and mounted into a 3D printed frame. This planar polarizer was placed on an azimuthal turntable at a fixed distance of approx. $4\,\mathrm{m}$ to the radar and then measured over its self-rotation. The corresponding measurement setup and the self-rotation process is shown in Fig. 5.4. In Fig. 5.5a, Pauli decomposed and RGB color coded results of a vertically aligned polarizer are shown. A trace of mostly magenta color is left over the rotation angle, having a bright peak at $90°$, where the polarizer front faces the radar sensor. Basically, the trace is left by the polarizer and its 3D printed frame. In order to determine the orientation of the polarizer, the ratio of HH/VV magnitudes is

Figure 5.4: Measurement setup with planar polarizer rotated around itself in azimuthal direction.

calculated and plotted in dB over the rotation angle in Fig. 5.5b. This dB-ratio is negative for a vertical polarizer and positive for a horizontal polarizer. As can be seen, the vertical polarizer can clearly be determined by means of this ratio. Another measurement with the same polarizer orientated in horizontal direction, was performed. The results can be observed in Fig. 5.6. Again, a magenta trace is left over the rotation angle, having a bright peak at approx. $90°$. Obviously, the orientation of the polarizer can not be determined by an evaluation of the squared magnitude of the Pauli components, Fig. 5.6a strongly resembles Fig. 5.5a. This becomes evident, if Eq. (2.14d) and the Pauli parameters in Eq. (2.19) are considered. To obtain the orientation, the HH/VV magnitude ratio has to be evaluated, which can be done by considering Fig. 5.5b. As can be seen, now the horizontal orientation of the polarizer can obviously be identified.

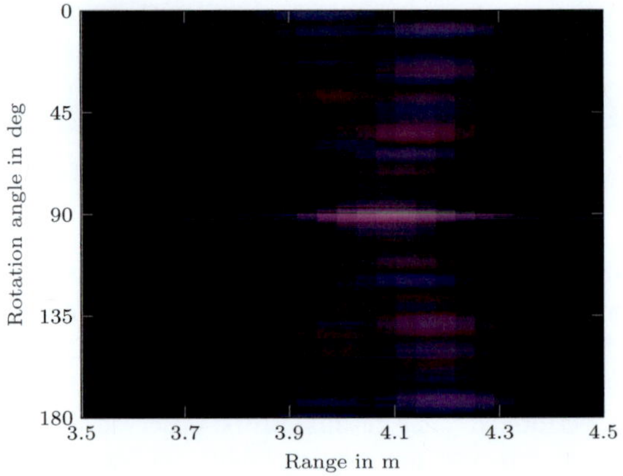

(a)

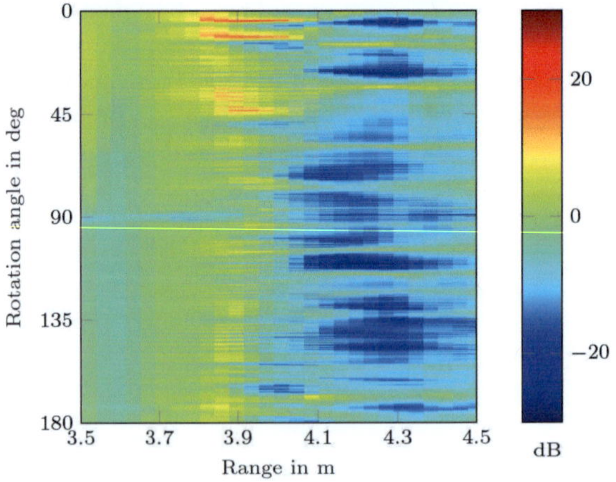

(b)

Figure 5.5: Measurement results of a planar polarizer, orientated in vertical direction and rotated in azimuthal direction. (a): Pauli decomposition RGB results. (b): HH/VV co-pol ratio in dB.

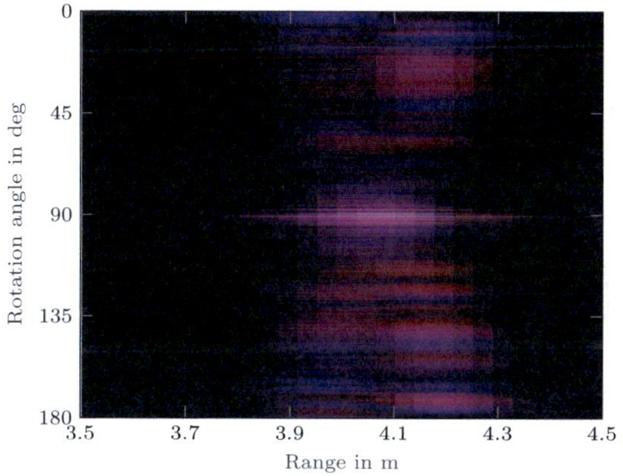

(a)

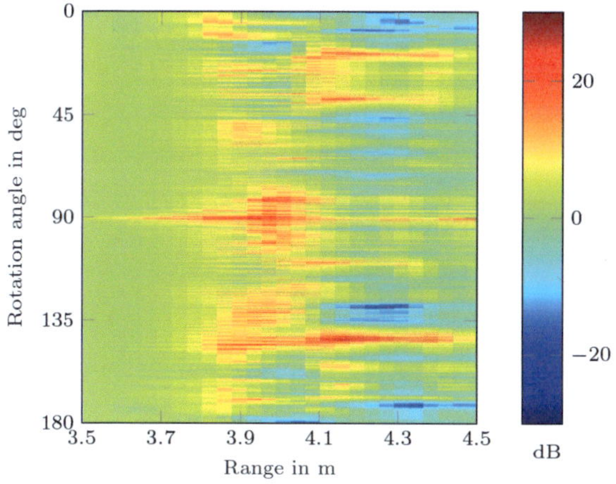

(b)

Figure 5.6: Measurement results of a planar polarizer, orientated in horizontal direction and rotated in azimuthal direction. (a): Pauli decomposition RGB results. (b): HH/VV co-pol ratio in dB.

5.3 Measurements of Targets on Rotary Table

Taking a step towards real targets occurring in traffic situations, especially weak targets like two-wheeled vehicles are interesting, because they generally exhibit small RCS values and are thus difficult to detect. In literature, e.g. in [BC15; BC16; Ulr08], RCS measurements of two-wheeled vehicles on rotary tables have already been performed, but - to the author's knowledge - no fully polarimetric measurements at MMW frequencies were published yet. The only work that was published so far in this direction deals with polarimetric micro-Doppler measurement results at X-band frequencies, presented in [Ryt+15].

5.3.1 Rotated Two-Wheeled Vehicles

Four different two-wheeled vehicles are selected for the polarimetric measurements - an off-road enduro motorcycle (type Yamaha WR450F) with a 21 inch tyre in the front and an 18 inch tyre in the back, a pedelec (from pedal electric cycle, type Bergamont E-Line C-XT Trapez) with a bicycle rack and 28 inch tyres, a child bike (type Puky Lillifee) with 12 inch tyres and a trekking bike (type Staiger New Orleans) with 28 inch tyres. Photos of the measured two-wheeled vehicles are shown in Figs. 5.7 and 5.8.

(a) (b)

Figure 5.7: Measured two-wheeled test vehicles on a turntable in an anechoic chamber on their start positions. (a): Trekking bicycle of type Staiger New Orleans. (b): Pedelec of type Bergamont E-Line C-XT Trapez.

(a)

(b)

Figure 5.8: Measured two-wheeled test vehicles on a turntable in an anechoic chamber on their start positions. (a): Off-road enduro motorcycle of type Yamaha WR450F. (b): Child bike of type Puky Lillifee.

The bicycles have an aluminum frame and bicycle locks wrapped around the seat posts, as can be seen in Figs. 5.7a and 5.8b. This selection of four vehicles is certainly just a small part of the huge range of two-wheeled vehicles. However, the aim was to pick out significant examples of this range of vehicles that represent the most important classes - motorbikes, standard bicycles, pedelecs and child bikes. In consideration of future classification purposes it is important to find out common features of two-wheeled vehicles. In this first approach, only yaw rotations of the objects are investigated and the vehicles are rotated clockwise around their center positions. The rotation start positions of the vehicles can be seen on the pictures in Fig. 5.7. Before Pauli-processed data is presented, the calibrated magnitudes of the S_{vv} component of the Yamaha motorbike measurement are shown exemplary for co-polar RCS in Fig. 5.9 over range and rotation angle. As can be observed in Fig. 5.9a, the classical VV component RCS, as would have been measured by a single-polarized radar in vertical direction, shows magnitudes up to a maximum of approx. $11.8\,\mathrm{dBsm}$, in good agreement with the results stated in [Ulr08]. As would have been expected, the exemplary cross-polar component RCS of S_{hv}, shown in Fig. 5.9b, has much smaller magnitudes with a maximum of only approx. $-4.5\,\mathrm{dBsm}$. In order to demonstrate the additional value of a polarimetric radar, the RGB images of the RCS measurement results of the motorbike and the pedelec after the coherent Pauli decomposition are shown in Fig. 5.10. Compared to the “classical" processing by only evaluating magnitudes, the Pauli decomposition also takes the relative phases between orthogonal polarized components into account. In contrast to the results shown in Fig. 5.9, the complete color-coded Pauli RCS in Fig. 5.10a contains much more information. Furthermore, in Figs. 5.10b, 5.11a and 5.11b the RCS RGB images of the pedelec, the trekking bike and the child bike are shown. A clearly distinct polarimetric signature of each measured object is obtained over range and rotation angle. It can be observed that the wheels of the vehicles show mainly odd bounce reflection (blue) behavior whereas the frame and engine of the vehicles (if present) exhibit relatively balanced contributions of all scattering processes together. Furthermore, the forks and handle-bars mainly reflect in an even bounce manner (red), while the rims and spokes mostly show very low reflection. Cross-polarized scattering (green) occurs primarily in oblique positions of the test vehicles to the radar LOS. Additionally, the range extension of the objects can clearly be detected, even if at some rotation angle steps shadowing of the parts behind dominant scattering parts exists. Given these characteristic polarimetric “fingerprints", the performance of modern classification methods might be enhanced, compared to single-polarized radar systems that lack polarimetric information.

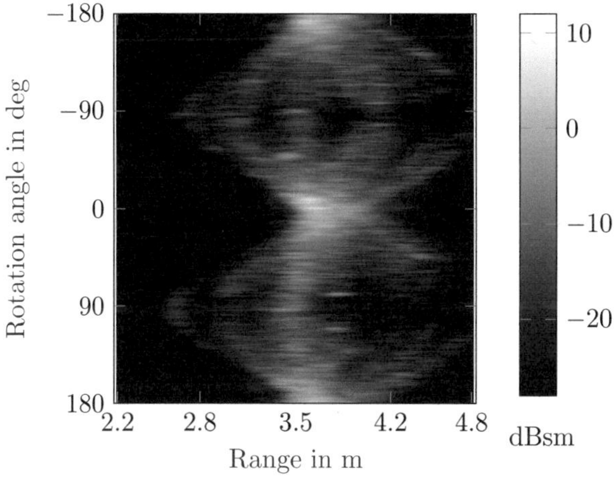

(a)

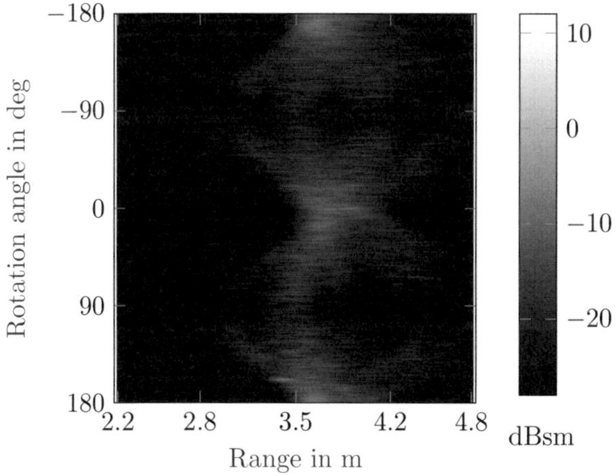

(b)

Figure 5.9: Conventional co- and x-pol RCS values of the measured Yamaha motorcycle in Fig. 5.8a over range and rotation angle. (a): RCS of the S_{vv} component. (b): RCS of the S_{hv} component.

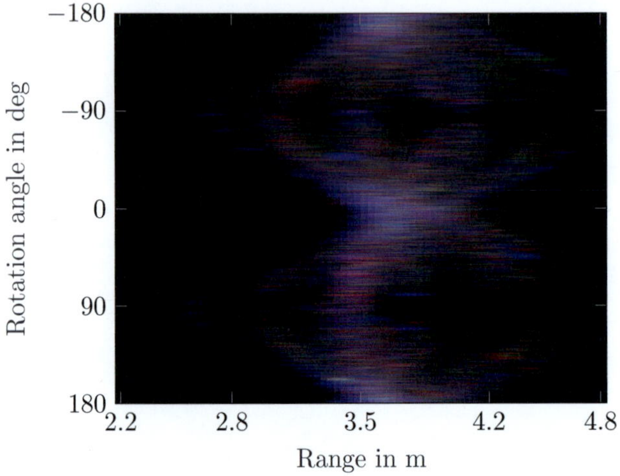

(a)

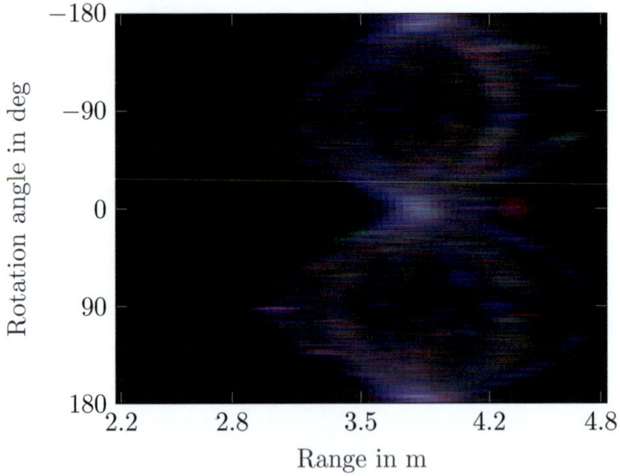

(b)

Figure 5.10: Color-coded RGB images of measured Pauli-decomposed polarimetric RCS values over range and rotation angle of (a): enduro motorbike shown in Fig. 5.8a and (b): pedelec shown in Fig. 5.7b.

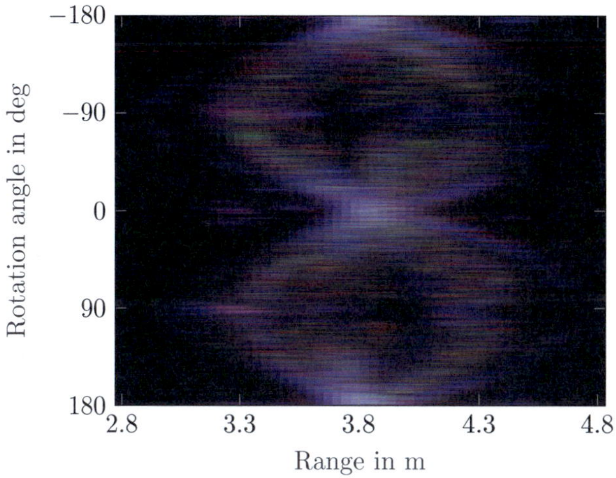

(a)

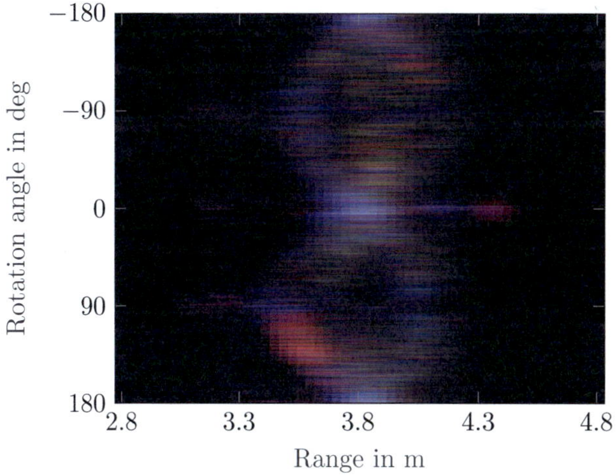

(b)

Figure 5.11: Color-coded RGB images of measured Pauli-decomposed polarimetric RCS values over range and rotation angle of (a): trekking bike shown in Fig. 5.7a and (b): child bike shown in Fig. 5.8b.

5.4 Analysis of Multipath and DOA Detection

Multipath effects in radar measurements can degrade the performance of DOA estimation by interference and can produce ghost targets in multipath directions that are mirror images of the real targets and are thus difficult to eliminate [Sko02]. If multipath reflections occur on a road, an interference pattern, dependent on the target's height, influences the received signal strength. The fading of interference minima can cause targets to disappear in certain distances [Die+11; Sko02]. Also, the signal phase is influenced by coherent multipath interference which complicates or even corrupts DOA estimation. If multipath reflections occur on a guardrail, the most probable effects are ghost targets or interference, depending on the specific scene, the vehicles shape and the radar's range resolution.
As was shown in the foregoing sections and chapters, polarimetric radars provide information about the parity of reflections such as even and odd bounce. This makes them promising candidates to reveal multipath reflections. Below, effects of polarimetric multipath propagations are discussed based on examples of multipath propagation over guardrails and roads. Furthermore, measurement results in an anechoic chamber demonstrate the parasitic interference effects of fading (which goes hand in hand with a corrupted DOA estimation) and ghost targets by multipath measurements of canonical objects. The change of polarimetric features in dependence of aspect angle is discussed. Also, measurements of the separability in angular direction between targets located in the same distance but with different polarimetric properties (as co- & x-pol or even & odd bounce), are presented. At the end of the section, a polarimetric filtering approach is introduced that allows polarimetric feature-selective DOA estimation.

5.4.1 Multipaths Over Roads and Guardrails

In Fig. 5.12, multipath propagations over guardrails and over roads, which are important cases in automotive radar applications, are presented. Since automotive radars operate in the E-band and typical structures in traffic sceneries are very large compared to MMWs, wave propagation according to geometrical optics (GO) can be assumed.

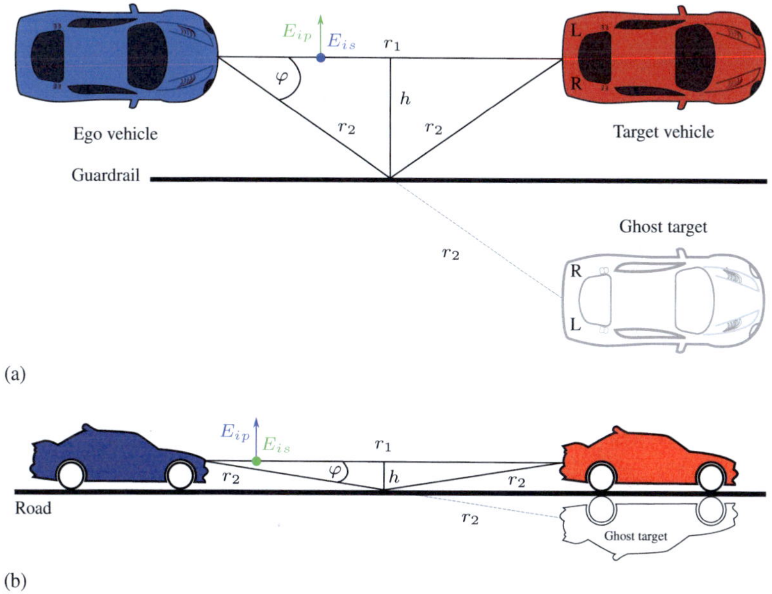

Figure 5.12: Two important scenes of multipath propagation in automotive environments. (a): Guardrail case. (b): Road case.

Multipaths Over Guardrails

Considering Fig. 5.12a, without loss of generality the vehicles and their scattering centers are located in the same distance h to the guardrail. Guardrails are metallic and the incidence can range from grazing to steep, depending on the distance h of ego and target vehicle to the guardrail. The main polarimetric effect occurring at guardrail reflections is that linear polarization is transformed to elliptical polarization upon reflection and vice versa. This change occurs due to a gradual phase shift between parallel (p)- and perpendicular (s)-polarizations to the guardrail (indicated by fields E_{is} and E_{ip} in Fig. 5.12a) over incidence angle. At a special incidence angle, called the principal angle, the phase shift between p- and s-polarizations becomes $90°$ and linear polarization is turned completely into circular polarization and vice versa. Attenuation effects are negligible for the guardrail case due to the metallic material and its high conductivity [BW99]. Rough surface and diffuse scattering gen-

erally increase incoherent backscattering and thus act in a depolarizing way. This depolarizing effect can be described quantitatively by the scattering entropy, cf. Section 2.6.2. In this section, rather smooth surfaces with a low scattering entropy are considered, so that a distinct classification in even and odd bounce is possible.
Three paths are important to analyze in case of Fig. 5.12a. At first, the direct path $d_{r1} = r_1$ is considered. The information on this path contains the direct backscatter and correct DOA information about the target. The first multipath is $d_{r12} = (r_1 + 2r_2)/2$. This path has one bounce (reflection) more than the direct path and thus changes an odd bounce scattering process into an even bounce one and vice versa. This path only exists for a spherical target or a target with strong surface curvature and/or diffuse scattering. Otherwise, the radar signal is not scattered in a different direction than the backscatter direction and the path d_{r12} is thus non-existent. The second multipath $d_{r2} = 2r_2$ describes the propagation with two more reflections than the direct path and thus the bounce information of even and odd bounce is maintained. Furthermore, the target is seen in a different aspect angle than from the direct path which leads to deviating polarimetric scatter information.
Two cases of problems arise with multipath reflections:

- The indirect path length lies within or close to the range bin of the direct path. The resulting coherent constructive and destructive interference with the direct path leads to fading effects on the Rx channels. As a result, the DOA estimation performance is deteriorated and in the worst case this leads to a wrong AOA estimation.

- The indirect path length lies in a different range bin as the direct path. The multipath reflection is detected as a ghost target with a ghost AOA φ. This ghost target is a mirror image of the real target and can only be revealed, if the mirror effect is detectable or by additional information processing (e.g. that the target velocity is unreasonable at the detected position). In the guardrail case of Fig. 5.12a this is especially difficult, because of the bilateral (left/right) symmetry of most vehicles.

These problems, if present, can deteriorate the otherwise high performance of automotive radars.

Multipaths Over Roads

When the radar signal is reflected from a road's surface, the road can be modeled in first approximation as a dielectric material. The reflection coefficients for dielectrics in the s- and p-polarized cases (indicated by E_{is} and E_{ip} in Fig. 5.12b) are described by the Fresnel equations. Ordinarily, the micro-roughness of asphalt is smaller or equal to one quarter of the free-space wavelength λ ($3.9\,\mathrm{mm}$) at $77\,\mathrm{GHz}$. Also, the statistical road unevenness leads to deviations from specular reflections. Therefore, modified Fresnel equations have to be used for reflections on roads [SDW00]. Furthermore, because of the roughness, the reflected wave is partly depolarized during the diffuse scattering process which increases the scattering entropy. Typically, radar sensors are not mounted on top of vehicles but rather close to the road. Thus, the incidence of over-the-road reflections is rather grazing. Then the direct path length d_{r1} is very close to the indirect path lengths d_{r12} and d_{r2}. Combined with the limited range separability, the indirect paths lie most probably in the range bin of the direct path and lead to interference effects and fading. Due to the rather grazing angle, the amplitude and phase differences of s- and p-polarized reflections are just marginal. For example, the reflectivity differences between s- and p-polarizations of a typical realistic rough asphalt surface for a monostatic sensor mounted at $30\,\mathrm{cm}$ height and a point target at the same height in a close distance of $5\,\mathrm{m}$ amounts to only $4.5\,\mathrm{dB}$ [SDW00]. Thus, the superposed power of all paths received by the radar follows an interference pattern over the target's range that is mostly independent of the polarization state [Die+11]. This gives rise to partly vanishing targets for single-polarized as well as fully polarimetric radars in certain target distances. An approach to tackle this issue would be to use a bistatic or quasi-monostatic radar with widely distributed antenna channels in order to make use of path differences between the receive channels. If these path differences become large enough, the fading will not occur on all channels equally. On the other hand, it is possible to estimate the height of objects on the basis of the interference pattern [Die+11].
Evaluating the foregoing statements, multipath reflections over roads show minor polarimetric detection potential and are thus not further investigated in this thesis.

5.4.2 Measurement Results

In the following, exemplary measurement results are presented in order to demonstrate the stated polarimetric effects in Section 5.4.1. For the DOA estimation, the two central VA MIMO blocks with overlapping HH & VV and HV & VH channels in Fig. 3.3 are evaluated using the method described in Section 2.3.2. The two gaps in the arrays at the edges are filled with zeros. Subsequent to the calibration from Section 4.1, the complete polarimetric S-matrix is formed. In order to differentiate the data in terms of polarimetric information, the measured S-matrix is then decomposed according to the coherent Pauli decomposition which was introduced in Section 2.6.1. As a new approach, the DOA estimation is performed for each component of the Pauli decomposition, where $\tilde{a}$ describes the odd bounce scattering part of the DOA spectrum, $\tilde{b}$ even bounce scattering, $\tilde{c}$ cross-polar scattering and $\tilde{d}$ antisymmetric components of the S-matrix which can mainly be neglected for a quasi-monostatic radar as is used in this work [LP09; Mar12]. The squared magnitude of the Pauli components describes the corresponding part of RCS. An overview of the described signal processing chain is presented in Fig. 5.13.

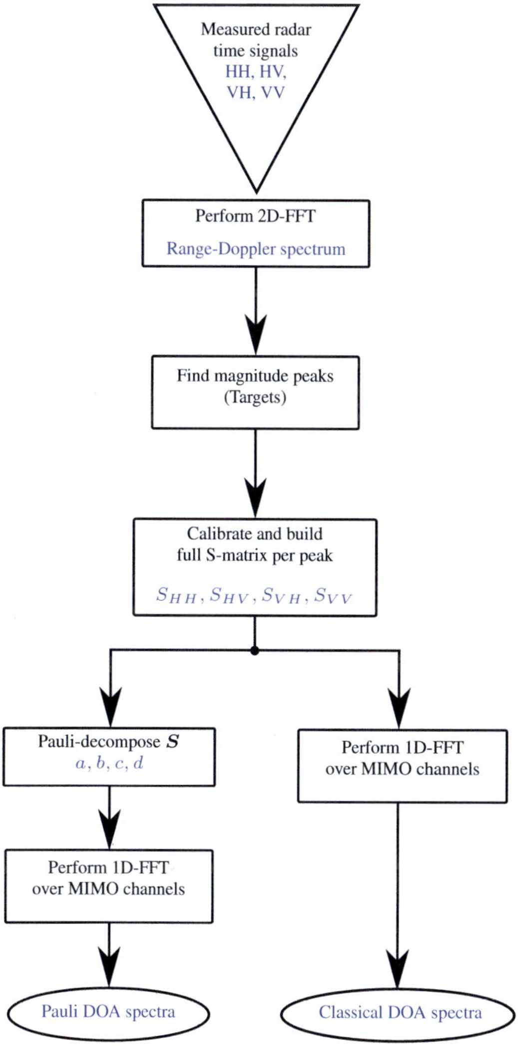

Figure 5.13: Flow chart of the signal processing chain used in this section.

Measurements of Polarimetric Multipath Effects

The performed measurements emulate the guardrail case from Fig. 5.12a. The corresponding setup is shown in Fig. 5.14. To model the guardrail case, the

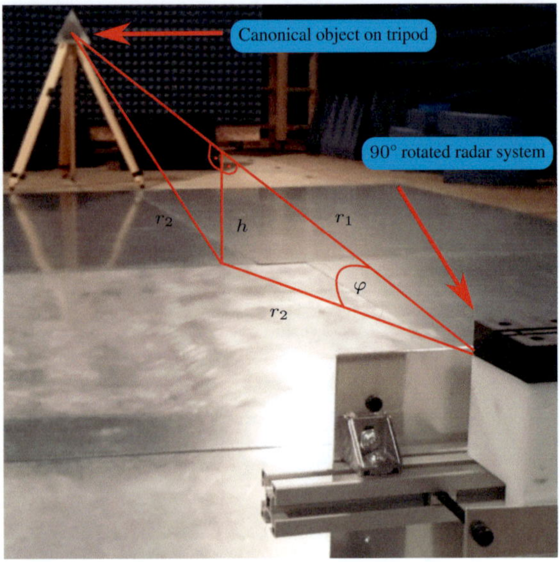

Figure 5.14: Measurement setup in an anechoic chamber for modeling the guardrail case of Fig. 5.12a with rotated radar sensor and trihedral.

floor between target and radar is covered with metallic sheets. Also, the radar sensor is rotated by 90° so that the azimuthal direction is pointing towards the ground (the azimuthal antenna pattern has a broad 3 dB-beamwidth of 60°, cf. Section 3.2.2). The distances are chosen in such a way that all three paths are separable - with a measurement bandwidth of $B = 2\,\mathrm{GHz}$ the physical range bin size is 7.5 cm [Hua+11]. The lengths in the setup amount to $d_{r1} = 5.2\,\mathrm{m}$, $h = 1.25\,\mathrm{m}$ and $d_{r2} = 2((r_1/2)^2 + h^2)^{1/2} \approx 5.8\,\mathrm{m}$. Consequently, $d_{r12} \approx 5.5\,\mathrm{m}$ and $\varphi = \cos^{-1}(r_1/2r_2) \approx 25.7°$. Different canonical objects are measured with these setup parameters.
At first, a metallic sphere with radius $r = 15\,\mathrm{cm}$ is installed as target. The uncalibrated measured magnitudes are incoherently integrated over each of the 4 overlapping MIMO quasi-ULAs and the results are shown in Fig. 5.15.

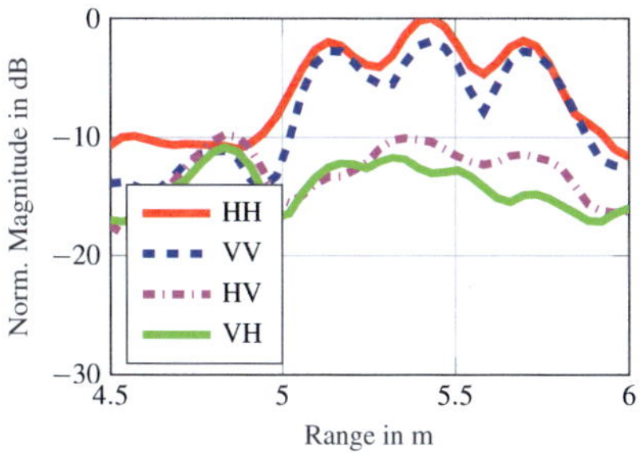

Figure 5.15: Uncalibrated range spectrum of sphere target measurement for all S-matrix components (incoherently integrated over all MIMO channels).

The three peaks of the different paths are clearly visible at $d_{r1} \approx 5.1\,\mathrm{m}$, $d_{r12} \approx 5.4\,\mathrm{m}$ and $d_{r2} \approx 5.7\,\mathrm{m}$, respectively. In the special case of a sphere, radiation is backscattered in many different angles due to its curved surface. Subsequent to the path selection and calibration, the DOA estimation, as stated in Section 2.3.2, is performed for each Pauli component. The estimation result for the direct path is given in Fig. 5.16a. The AOA is estimated correctly for the $\tilde{a}$ component, indicating that an odd bounce target is detected close to $0°$ incidence. For the indirect paths d_{r12} and d_{r2}, this is not the case. The corresponding results are shown in Figs. 5.16b and 5.16c. As can be seen, the DOA can not be estimated in the case of path d_{r12}. From d_{r1} to d_{r12} the RCS of the $\tilde{a}$ component changes from $-11\,\mathrm{dBsm}$ to $-18.9\,\mathrm{dBsm}$, whereas the RCS of the $\tilde{b}$ component changes from $-26.5\,\mathrm{dBsm}$ to $-7.4\,\mathrm{dBsm}$, illustrating the change from odd bounce to even bounce due to the extra reflection on the metallic ground. Nevertheless, because of the coherent superposition of many paths similar to d_{r12}, the DOA estimation is corrupted. Contrary to d_{r2}, where only very few rays are reflected by the metallic floor at such an angle that they hit the sphere to be exactly bounced back, in case of d_{r12}, many rays in different angles are bounced back by the sphere and are reflected by the metallic floor to hit the radar. Considering Eq. (2.12), a superposed wave E_{sp}

of the direct path wave E_{dp} and the indirect path wave E_{idp} at the nth antenna for a $\lambda/2$ ULA can be formulated as:

$$E_{sp,n} = E_{dp,n} + E_{idp,n} = E_{0r}\left(e^{jnk_1} + e^{jnk_2}\right) \tag{5.1}$$

$$E_{sp,n} = 2E_{0r}\cos\left(\frac{n(k_1 - k_2)}{2}\right)e^{j\left(\frac{n(k_1+k_2)}{2}\right)}, \tag{5.2}$$

with $k_1 = \pi\sin(\varphi_{dp})$ and $k_2 = \pi\sin(\varphi_{idp})$. From Eq. (5.2) it is obvious that the interference modulates the received amplitude as well as the phase. In the measurements, this causes a RCS fluctuation of up to $15\,\mathrm{dB}$ over the co-polar MIMO channels and makes it impossible to estimate a reasonable DOA. This effect can be observed in Fig. 5.17, where the RCS values are plotted per virtual MIMO channel position and polarization combination.

It is to be noted that the sphere is a pure co-polar target and thus the RCS values of the cross-polar combinations HV & VH are lower than the co-polar ones (HH & VV). The third multipath d_{r2} leads to the detection of a ghost target with a high odd bounce component at an AOA of approx. $26°$. The corresponding DOA estimation result is shown in Fig. 5.16c.

Next, a trihedral corner reflector with $18\,\mathrm{cm}$ edge lengths as in Fig. 5.14 is installed. The evaluation of the direct path at $5.2\,\mathrm{m}$ results in dominant Pauli RCS components of $|\tilde{a}|^2 \approx 18.8\,\mathrm{dBsm}$ and $|\tilde{b}|^2 \approx 1.3\,\mathrm{dBsm}$. The other components are neglectable. This clearly shows that the trihedral is classified correctly as a target with a high odd bounce component and a $17.5\,\mathrm{dB}$ lower even bounce component. The indirect path d_{r12} can not be detected in the measurement. This is because the trihedral is mostly a backscatter object and does not scatter radiation to other directions [Kno12]. However, the indirect path d_{r2} is detected at $5.8\,\mathrm{m}$. The Pauli decomposition for this path results in the dominant parts $|\tilde{a}|^2 \approx 4\,\mathrm{dBsm}$ and $|\tilde{b}|^2 \approx 1\,\mathrm{dBsm}$. Also, the DOA estimation of $\tilde{a}$ and $\tilde{b}$ leads to $\varphi \approx 27°$ as expected. This shows that the multipath is detected as a ghost target at an AOA of 27° with an even bounce RCS part that is only $3\,\mathrm{dB}$ weaker than the odd bounce part. That is due to the high aspect angle, in which the trihedral is illuminated by the incoming rays. A trihedral seen in this aspect angle is not a pure trihedral anymore and exhibits a certain amount of dihedral scattering [Kno12]. A part of scattered power comes from the upper two edges of the trihedral. This can be visualized, if the trihedral in Fig. 5.14 is observed from the direction of ray r_2.

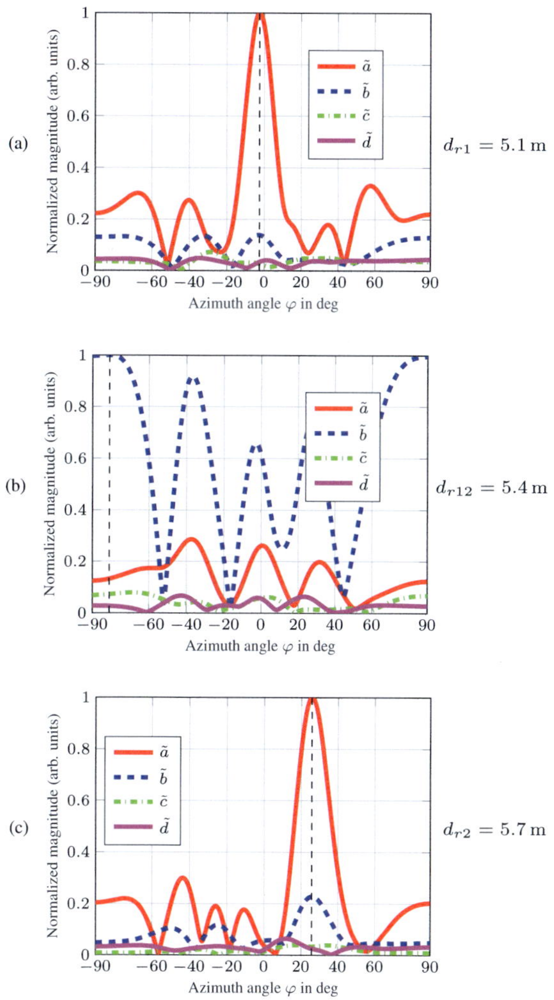

Figure 5.16: Spatial spectra of Pauli components over azimuth angle of the measurement shown in Fig. 5.14 with a sphere instead of a trihedral. (a): Direct path. (b): Indirect path d_{r12}. (c) Indirect path d_{r2}.

The same aspect angle effect is observed when the trihedral is exchanged by a dihedral corner reflector with edge lenghts of 20 cm and its seam rotated by 0°. For the direct path, the dominant two Pauli RCSs are $|\tilde{a}|^2 \approx 7.1\,\text{dBsm}$ and $|\tilde{b}|^2 \approx 24.6\,\text{dBsm}$ and for the indirect path d_{r2} (DOA also estimated at $\varphi \approx 27°$) $|\tilde{a}|^2 \approx 7.9\,\text{dBsm}$ and $|\tilde{b}|^2 \approx 19.6\,\text{dBsm}$. This shows again, that the targets change if seen in a different aspect angle.
For a metallic plate with edge lengths of 28 cm however, only the direct path is measurable, because of its very weak scattering in off-broadside directions [Kno12].

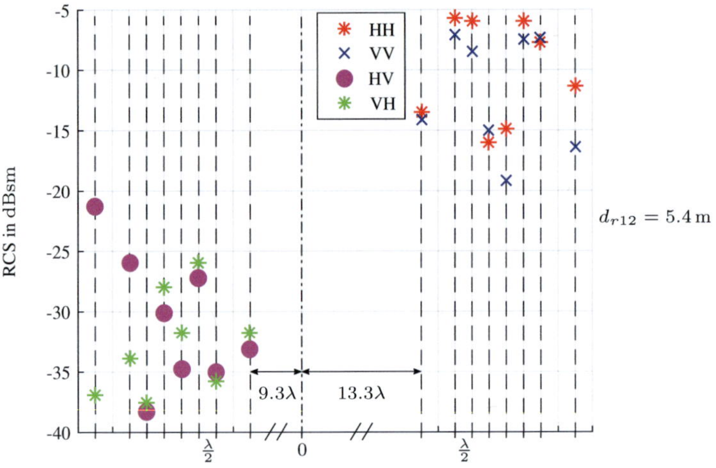

Figure 5.17: Measured RCS of sphere, plotted on location of virtual MIMO channels over array element position for indirect path d_{r12}. Dashed lines indicate the virtual array element locations and the distances between them are given in units of one wavelength on the abscissa.

Angular Separability of Different Canonical Objects

A validation of the polarimetric DOA estimation performance is presented in the following measurement. A trihedral and a dihedral (rotated once by 0° and another time by 45°) are arranged in the anechoic chamber at approx. the same radial distance to the radar, but in different off-broadside directions. A photo of the setup is given in Fig. 5.18.

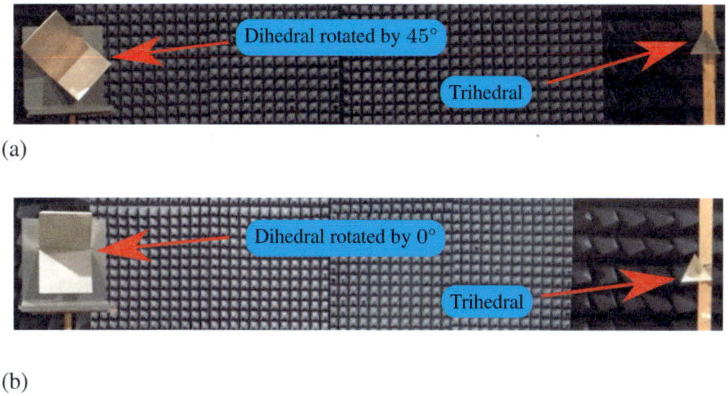

Figure 5.18: Measurement setup of canonical targets separated only by angular position in anechoic chamber. (a): Dihedral as x-pol target and trihedral as odd bounce target. (b): Dihedral as even bounce target and trihedral as odd bounce target.

The uncalibrated magnitudes of the range spectrum are again incoherently integrated over each overlapping MIMO ULA and the results are shown in Fig. 5.19.

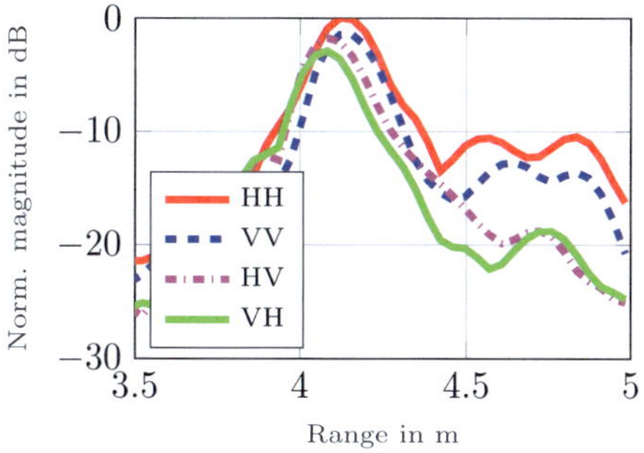

Figure 5.19: Uncalibrated range spectrum of the measurement from Fig. 5.18a for all S-matrix components (incoherently integrated over all MIMO channels).

The range spectrum peaks of the targets are overlapping at about 4.1 m and are thus not clearly separable. An evaluation of the Pauli component based DOA estimation for the overlapping range peaks is shown exemplary for the dihedral rotated by 45° case (which is a pure cross-polar target [3]) in Fig. 5.20.

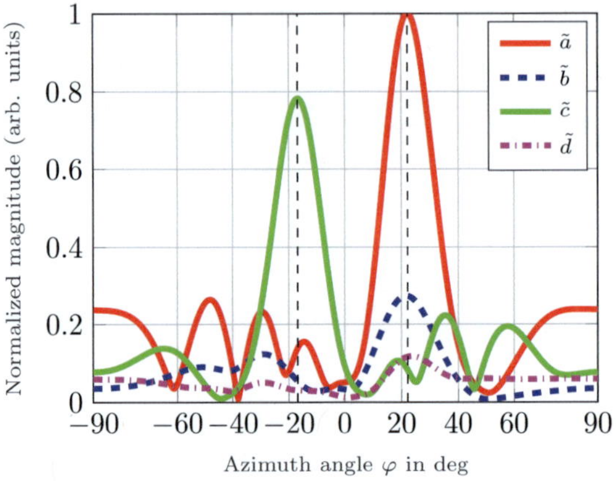

Figure 5.20: Spatial spectrum of Pauli components over azimuth angle of the measurement from Fig. 5.18a.

As observed, the AOA of $\tilde{c}$ shows an incidence angle of about $-18°$, whereas the AOA of $\tilde{a}$ shows an incidence angle of about $22°$. The targets can be clearly identified by the corresponding Pauli components and the AOA is estimated correctly. The same holds, if the dihedral is rotated to $0°$ (which is a pure co-polar even bounce target [3]). Here, the AOA of the even bounce $\tilde{b}$ component shows the dihedral at about $-17°$ and the trihedral is still detected at about $22°$ of the odd bounce $\tilde{a}$ component.

Polarimetric Multipath Filtering Based on Pauli Parameters

An interesting case to investigate is the one when direct and indirect paths (e.g. object and ghost target) are overlapping in one range bin, resulting in one peak in the range spectrum but in two separate peaks in the DOA spectrum. Then, it cannot be decided which one of the peaks belongs to the real target and which one to the ghost. An experimental study was thus conducted in order to test if the polarimetric features could help in this case to filter out the ghost target's peak from the DOA spectrum by using a signal superposition.
So far, only measurements of objects positioned in boresight of the radar were shown. In order to reveal the polarimetric multipath effects also in off-broadside scenarios, the trihedral from Fig. 5.18 was placed in a position of approx. $-25°$ at $5.2\,\mathrm{m}$ radial distance to the radar. A plot of the measured

DOA spectrum is shown in Fig. 5.21a. The AOA of the trihedral is correctly estimated at about $-26°$. As can be observed, the object only exhibits a co-polar DOA spectrum. As already stated in Section 5.4.2, the trihedral is an object that exhibits only pure backscatter, due to its faceted geometry. Thus, the indirect path d_{r12} is not detectable. It follows that only the indirect path d_{r2}, which generates a ghost target, is found. The range resolution of the radar used in this chapter is high enough ($7.5\,\mathrm{cm}$) to separate direct and indirect paths d_{r1} and d_{r2} due to the distance between them ($60\,\mathrm{cm}$ with the measurement parameters stated in Section 5.4.2). As a result, it is difficult to arrange a scenario in the rather small anechoic chamber in such a way that the different path lengths d_{r1} and d_{r2} come closer together so that they overlap in one range bin. However, a filtering ability can only be demonstrated if direct and indirect paths overlap or if they can not be separated anymore. This would be the case for a radar with lower range resolution, so that for the measurement setup used in this chapter direct and indirect paths overlap in one large range bin. For example, a radar with range resolution of $60\,\mathrm{cm}$, resp. with a bandwidth of $250\,\mathrm{MHz}$, could be used. Since the bandwidth of the radar used in this chapter can not be adjusted accordingly, a different approach must be used. Thus, the complex values of the peak from the indirect path measurement of the boresight positioned trihedral from Fig. 5.14 are extracted and superposed to the complex values of the peak from the direct path measurement results of the off-broadside trihedral. The superposition is done in the range spectrum and is placed on the same range bin as the peak of the detected direct path measurement. In other words, the indirect path of the measurement from Fig. 5.14, which is detected at a distance of $5.8\,\mathrm{m}$, is superposed to the direct path of the measurement of the off-broadside trihedral, which is detected at a distance of $5.2\,\mathrm{m}$. Additionally, to complicate the detection of the direct path at $5.2\,\mathrm{m}$, the strength of the indirect path is amplified by a factor of six. Thus, the superposed spectrum comprises one time the direct path and six times the indirect path at the range bin of the direct path. This approach produces distinct fading over the MIMO channels. The resulting DOA spectrum is presented in Fig. 5.21b. As can be seen, the peak of the indirect path is dominant over the one of the direct path in the DOA spectrum. If the highest peak would be considered as the AOA of the target, a significant error would be made. Also, the spectrum could be misinterpreted as a two-target scenario. By utilizing the Pauli parameters, it becomes possible to separate the spectral parts of the DOA spectrum concerning their different polarimetric features. In Fig. 5.21c, the DOA spectra of all Pauli components are plotted over azimuth

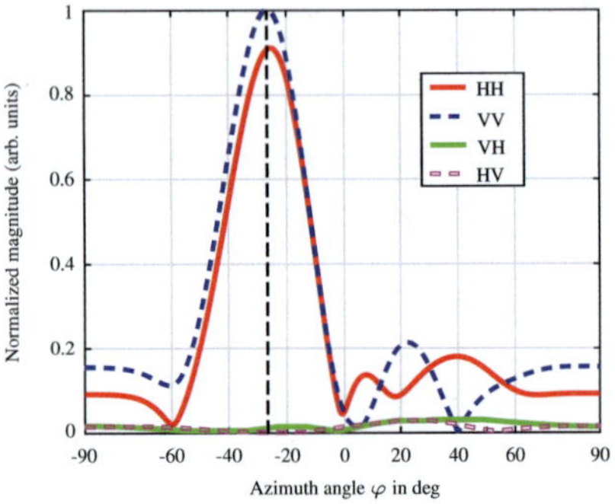

(a)

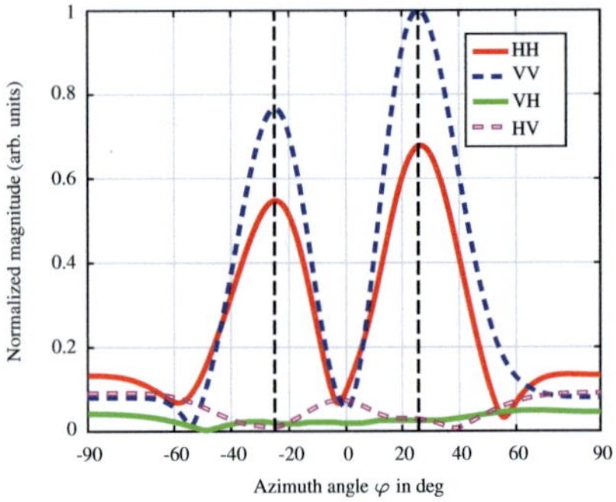

(b)

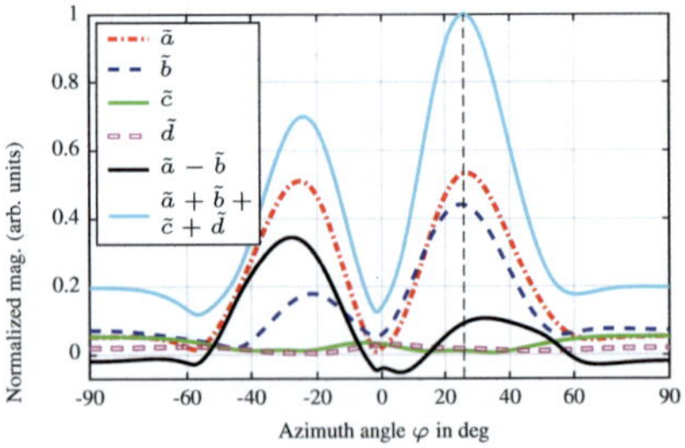

(c)

Figure 5.21: Measured spatial spectra over azimuth angle of a trihedral placed at approx. $-25°$ off-broadside and $5.2\,\mathrm{m}$ distance. (a): Direct path. (b): Superposition of $1\times$ (a) and $6\times$ indirect path d_{r2} from the measurement of Fig. 5.14. (c) Pauli parameters of (b) with a superposition and a difference of them.

angle with several superpositions of them. These superpositions allow a selective filtering of polarimetric parts of the DOA spectra. For example, the spectrum of $\tilde{a} - \tilde{b}$ allows a DOA estimation of the pure odd bounce part with a negative superposition of any even bounce part of the S-matrix of the considered target. Using this superposition, the correct AOA of the trihedral can be estimated at about $-27°$. This is based mostly on the aspect angle dependence effect which was introduced in Section 5.4.2. The trihedral seen from the indirect path d_{r2} exhibits a high even bounce scattering part. Furthermore, the superposition of all parts $\tilde{a} + \tilde{b} + \tilde{c} + \tilde{d}$ of the DOA spectrum shows the combined DOA spectrum of the entire S-matrix of the detected target. It is used for normalization of the given graphs.
In order to use the filtering advantageously, the correct difference of Pauli components has to be selected. For example, the difference of $\tilde{a} - \tilde{c}$ in Fig. 5.21c would have almost no effect and would result in just the odd bounce component $\tilde{a}$, because the cross-polar component $\tilde{c}$ is almost zero. One approach is to use a threshold value in the spectra on which a decision can be based, if a difference is feasible or not. In this example, a threshold of 0.15 can be used. This would lead to the knowledge that parts $\tilde{a}$ and $\tilde{b}$ lie above the threshold value, whereas parts $\tilde{c}$ and $\tilde{d}$ remain below the threshold value. As a result, the difference of the $\tilde{a}$ and $\tilde{b}$ parts can be calculated and is promising to give new insights. However, it remains an open question if the detected polarimetric parts in the spectra originate from multipaths or from real objects in the same range bin, as in Fig. 5.20, if the measured scenario is unknown. Nevertheless, the approach enables the possibility to split the DOA spectrum in its polarimetric parts and make use of the filtered spectra. Detection of fading over the channels or other superior knowledge like taking into account the shape of the target or also detection of a close-by guardrail is needed to reveal the physical origin of the spectral parts.

6 Classification of Polarimetric Measurement Data Based on Deep Learning

Getting back to the results of Section 5.3.1, the question arises how far classification methods based on the measured polarimetric "fingerprints" or generally polarimetric measurement data can lead to success. Due to their outstanding performance in image recognition, the application of CNNs for classification of polarimetric radar images is therefore investigated in this chapter. In order to assess this method on a fundamental level, several artificial radar targets that differ in physical dimensions and also in their polarimetric properties are designed and constructed. These targets are mounted on a rotary table in an anechoic chamber and are rotated in front of the fully polarimetric radar, in the same way as the targets in Section 5.3. For each rotation angle step, the targets are measured, where all measurements are taken at several distances. Then, a CNN is used to classify the measurement results into one of three superordinate classes of the constructed radar targets, each containing two different subordinate targets (sub-targets). Focus is lied on class decisions and not identification of specific sub-targets. A decision is taken for each single rotation angle step measurement and also for a majority vote on five consecutive rotation angle step measurements. All measurements presented in this chapter were performed in an anechoic chamber using the fully polarimetric radar sensor from Chapter 3 with the antenna from Section 3.2 that was calibrated by the method presented in Section 4.1. The utilized FMCW modulation parameters are listed in Appendix A.2. Parts of this chapter were published in [6].

6.1 Convolutional Neural Networks (CNNs)

CNNs incorporate convolutional layers that base on the multilayer perceptron (MLP), but use shared weights to achieve a convolutional effect [GBC16]. Therefore, they exploit data structures which are present in images so that learned features can be recognized in different parts of the entire image. The shared weights of the convolutional layers are limited in spatial dimensions by their receptive fields, similar to targets in radar images, which also occupy only a part of the entire image. This means that the total number of weights can be greatly reduced, allowing for deeper architectures and making the training process much more efficient in comparison to MLPs [GBC16]. In contrast to CNNs, MLPs comprise only fully connected layers with a lot more weights that have to be trained.

6.2 CNN Architecture and Training Process

The CNN architecture used in this chapter is a modified version of a CNN that performs well on the classification of handwritten digits from the modified national institute of standards and technology (MNIST) dataset [Lec+98]. Its architecture is depicted in Fig. 6.1. It consists of three convolutional layers

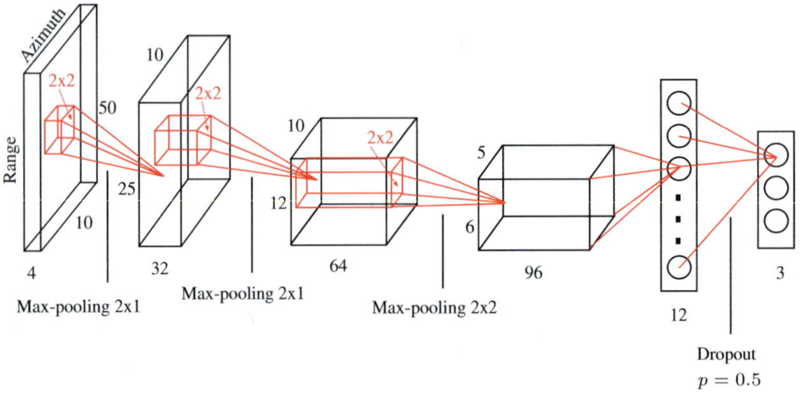

Figure 6.1: Schematic architecture of the used CNN with input and output data dimensions.

with rectified linear unit (ReLU) activations to enable learning of non-linear features [NH10]. Each of them is followed by a max pooling layer to condense the information and allow for more convolutional filters in the successive layers, e.g. 32 filters in the first layer, 64 in the second, etc. Because it is expected to be more difficult to distinguish similar radar targets that additionally vary much more with change of aspect angle than the handwritten digits, more filters (more capacity) than in the MNIST example are introduced. To improve generalization, a dropout with probability $p = 0.5$ is used in the last layer (fully connected layer). Together with the early stopping method, the dropout serves as regularization technique [GBC16]. Training is conducted using the stochastic gradient descent (SGD) method [GBC16] with a random start value of all used shared filter weights. In the training phase, a Keras implementation of the adaptive moment estimation (Adam) optimizer from [KB14] is used for adjusting the learning rates of the SGD method. The used loss function is cross-entropy loss, which is applied to the softmax output of the last layer (the second fully connected layer) [GBC16]. After training of 100 epochs, the accuracy on validation data for each epoch is used as an indicator at which epoch the training is stopped according to the early stopping approach. Accuracy is a performance metric that relates the examples which the model predicts correctly to the total number of examples [GBC16]. The resulting weights of this checkpoint are taken for the final CNN model.

6.3 Polarimetric Targets for Classification

In order to test the classification performance of machine learning algorithms, it is necessary to have a set of targets which are separable in polarimetric radar images, and can be reproduced reliably. For static measurements, the distinction between individual targets can be made e.g. by spatial extent, polarimetric RCS or spatial distribution of polarimetric RCS. The targets used in this chapter comprise canonical objects in form of dihedrals and boxes with a cross-section of $25\,\mathrm{cm} \times 25\,\mathrm{cm}$ that can be individually rotated around their axes in fixed steps. These canonical objects represent different polarimetric information, cf. Section 2.5. If orientated to the radar LOS, dihedral objects exhibit pure co-pol and even bounce scattering if rotated by 0 or $90°$ as well as pure x-pol scattering if rotated by $45°$. In the same orientation to the radar LOS, box objects exhibit pure co-pol and odd bounce scattering, independent of their rotation. Furthermore, a metallic frame with 4 bars of $1\,\mathrm{m}$ length is constructed. On each bar, one canonical object can be mounted in rotations of 0, 45 and $90°$ around the bar. A representative construction with several canonical objects mounted on the frame bars is shown in Fig. 6.2. This frame, together with a specific configuration of the objects on its bars, represents a specific target. For the measurements, it is set up on wooden trestles on a

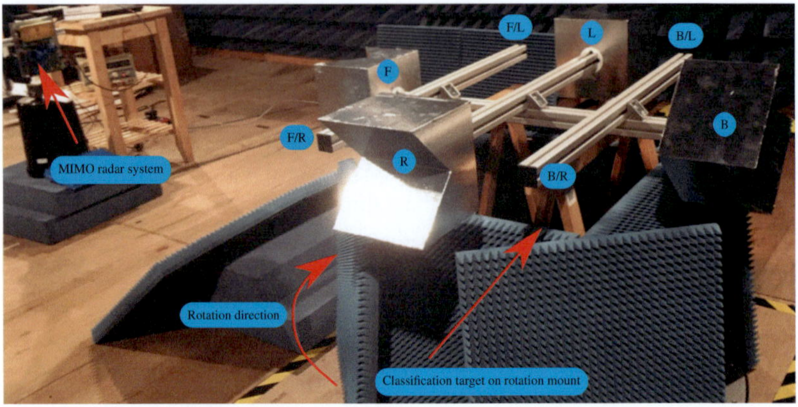

Figure 6.2: Measurement setup with metallic frame placed on wooden trestles above a rotary table. The metallic frame provides 8 locations to install canonical objects. In the photo, an exemplary target is shown with 4 canonical objects and their labeled positions.

Label	Canonical object	Placement
C1O1	Dih45°/Dih90°	F/B
C1O2	Dih-45°/Dih0°	F/B
C2O1	Dih45°/Dih90°/Dih0°/Dih90°	F/B/L/R
C2O2	Dih45°/Dih90°/Dih90°/Dih0°	F/B/L/R
C3O1	Dih45°/Dih90°/Dih0°/Dih90°/Box0°/Box45°	F/B/FL/FR/BL/BR
C3O2	Dih45°/Dih90°/Box0°/Box45°/Dih0°/Dih90°	F/B/FL/FR/BL/BR

Table 6.1: List of measured target classes and placement of their canonical objects with labels. Nomenclature: front (F), back (B), left (L), right (r).

rotary table in an anechoic chamber, like the two-wheeled vehicles in Section 5.3.1. The trestles are covered by MMW absorbers to reduce clutter. In total, 6 targets are built and measured. Their compositions are listed in Table 6.1. The targets are divided into three different classes (C1, C2, and C3), based on the number of mounted objects, their polarimetric diversity and the resulting physical dimensions. It can be seen that different classes have a portion of distinctly differentiating features, but also similar features, depending on the rotation position of the target on the rotary table.

6.4 Measurements and Data Preprocessing

Two kinds of measurements are performed in varying distances of target and radar, where each target is rotated around its axis to measure it in different aspect angles. For the training data, each target is rotated on the rotary table from $0°$ to $180°$ in $0.1°$ steps. A second measurement is taken in steps of $2°$, totaling 90 steps. This second measurement data set is split to build a validation dataset (comprising 19 rotation angle steps) and a test dataset (comprising 72 rotation angle steps), in order to validate the performance of the CNN. Training data measurements are only taken in a short distance of approx. $2.2\,\mathrm{m}$, whereas all other measurements are taken also in additional distances of $3.2\,\mathrm{m}$, $4.2\,\mathrm{m}$ and $5.2\,\mathrm{m}$. The training and classification process is performed on preprocessed data. Raw data from the radar is first processed by a 2D FFT which gives a range-velocity map. Because of the static measurements for each rotation angle step, the velocity dimension is neglected. Then, the FFT-processed data is calibrated by the method described in Section 4.1. Subsequent to these steps, the Pauli decomposition is used to decompose the polarimetric measurement data into co-pol odd bounce (Pauli a), even bounce (Pauli b) and x-pol (Pauli c) parts, as in Section 2.6.1. In order to attain azimuthal information about the objects, a further FFT is performed over the spatial dimension of the quasi-ULA antenna elements for each Pauli component, as in Section 5.4.2. Thus, the DOA spectrum for each Pauli component can be calculated independently.

6.5 Training and Classification Results

For the CNN training, absolute values of all four Pauli components are computed, since this gives a representation of the specific parts of polarimetric RCS in the data (cf. Eq. (2.20)). Classification is first analyzed by completely neglecting spatial information. This is done by leaving out calculation of the DOA spectrum and summing up all Pauli components over the range dimension (only within a region of $1.9\,\mathrm{m}$ around the target location), which leads to a pointwise polarimetric presentation of each rotation angle step measurement over the Pauli components. A plot of these data points is shown in Fig. 6.3 over normalized Pauli components. Normalization is performed for

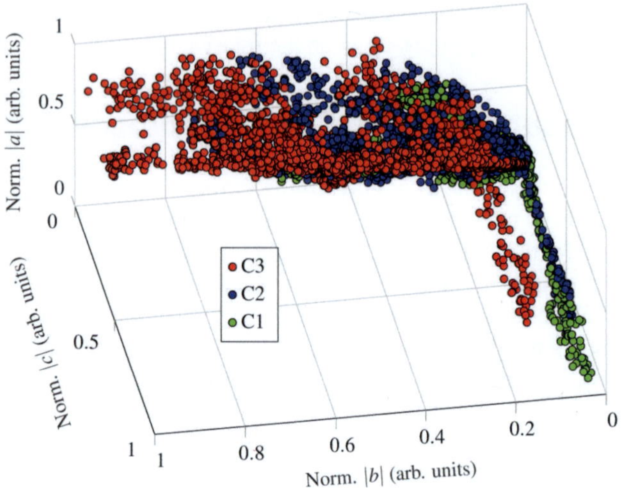

Figure 6.3: Measured Pauli components of all targets from Table 6.1 at a distance of 2.2 m. Components are summed up over the range dimension, normalized and plotted over Pauli component axes. Each dot denotes a single rotation angle step from 0° to 180°.

each Pauli component separately and taking into account the overall maximum of all points. Fig. 6.3 shows results of training measurements in 2.2 m distance for all three classes of targets with 1801 rotation steps between 0° and 180°. It can be observed, that a simple distinction of classes is only possible on the Pauli $|b|$-axis if the targets exhibit a non-negligible part of x-pol RCS on the Pauli $|c|$-axis. However, this classification could possibly just roughly decide between class C3 and the other two classes (C1 and C2) for a small number of aspect angles. All other points are mainly overlapping and no definite border for classification can be drawn.

Thus, also spatial information in range and azimuth is allowed for the CNN training in order to take advantage of the full classification potential. The first three Pauli components can be used to create color-coded RGB images, as described in Section 5.2. In the RGB code, the color red is dedicated to even bounce scattering RCS from the second Pauli component, the color green denotes the x-pol RCS from the third Pauli component and the color blue denotes odd bounce RCS from the first Pauli component. The small part of RCS from the fourth Pauli component is neglected for this color format, but it is taken into account for the CNN training. As for the whole dataset, dB values of the Pauli parameters are computed and a threshold of $-40\,\mathrm{dB}$ is introduced in the images, i.e. all values of the Pauli components smaller than this threshold are set to the threshold value (cf. Fig. 5.2). An exemplary RGB image of the color-

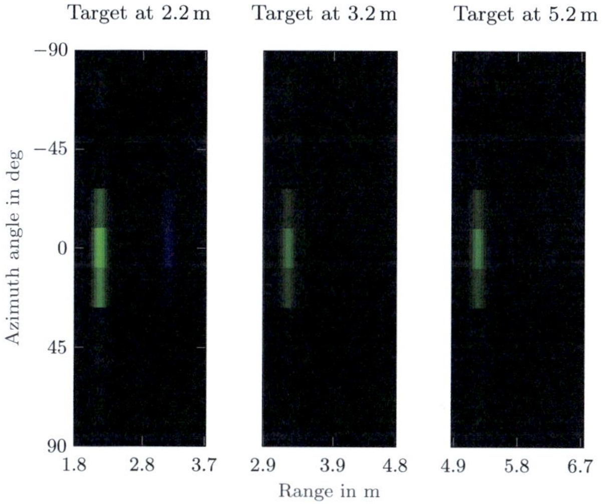

Figure 6.4: Pauli RGB color-coded images over range and azimuth dimensions of target C2O2 from Table 6.1 in three different measurement distances of 2.2 m, 3.2 m and 5.2 m. Images are created at 0° rotation angle of the target and normalized to the overall maximum of the entire dataset.

coded range-azimuth dataset of target C2O2 from Table 6.1 for 0° rotation is presented for three distances in Fig. 6.4. Each image is globally normalized to the maximum of the three Pauli components taken from the entire dataset. A dominant x-pol object can be observed. As listed in Table 6.1, targets C1O1 and C2O2 comprise a dihedral rotated by $\pm 45°$ in the front position. At a rotation angle of 0°, both targets are almost equal and the classification task is still extremely difficult. According to the dimensions of the targets measured in this chapter, they occupy max. 50 range bins ($1.9\,\mathrm{m}$) in range direction. In azimuthal direction only 10 azimuth bins are available, due to the number of quasi-ULA antenna elements and gaps, cf. Fig. 3.3 and Section 5.4.2. Because of the small number of azimuth bins, max pooling is greatly reduced in this dimension to avoid losing too much information. Training, validation, and test data sets are built by cutting out the specific portion of 50×10 range-azimuth cells of the whole radar spectrum that contains the targets, as in Fig. 6.4. This gives the radar images for the datasets. The training data set consists of all 1801 rotation measurements from $0°$ to $180°$ in a distance of $2.2\,\mathrm{m}$. It includes all targets from the three classes. Taking together all four absolute values of the Pauli parameters, this gives $50 \times 10 \times 4$ input values for the first convolutional layer of the CNN in Fig. 6.1. Since each of the Pauli com-

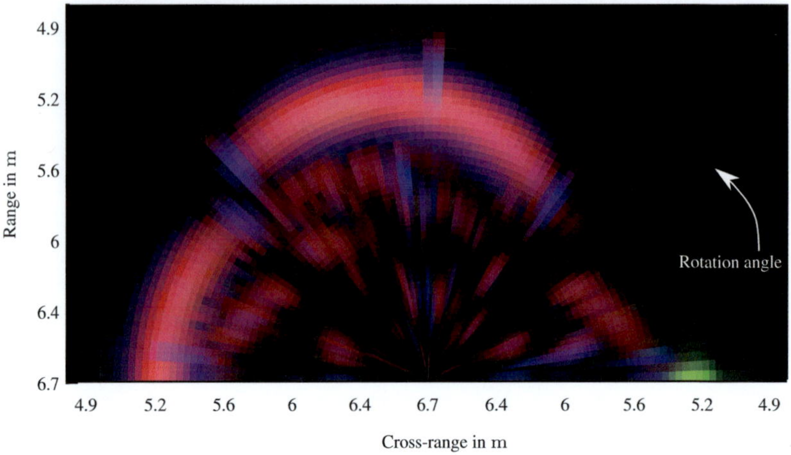

Figure 6.5: Measured Pauli RGB color-coded polar plot of C3O2 from Table 6.1 over range and cross-range for all rotation angle steps in a distance of $5.1\,\mathrm{m}$.

ponents describes a part of the polarimetric RCS, physical range-dependent effects of the RCS play a significant role. These effects range from near-field effects over interference effects (glint, scintillation, cf. Section 2.4) to a power-range dependency of the radar channel that is not exactly proportional to $1/r^4$ like in the free space propagation channel assumption with which the radar system was calibrated (cf. Section 4.1) [Kno12]. Therefore, several normalization approaches were tested regarding their classification accuracies. These approaches are discussed later on in Section 6.6. At first though, a global normalization, where each radar image is normalized to the overall maximum of the Pauli parameters obtained from the entire dataset, is introduced. By this method, RCS-variations in different distances are preserved.

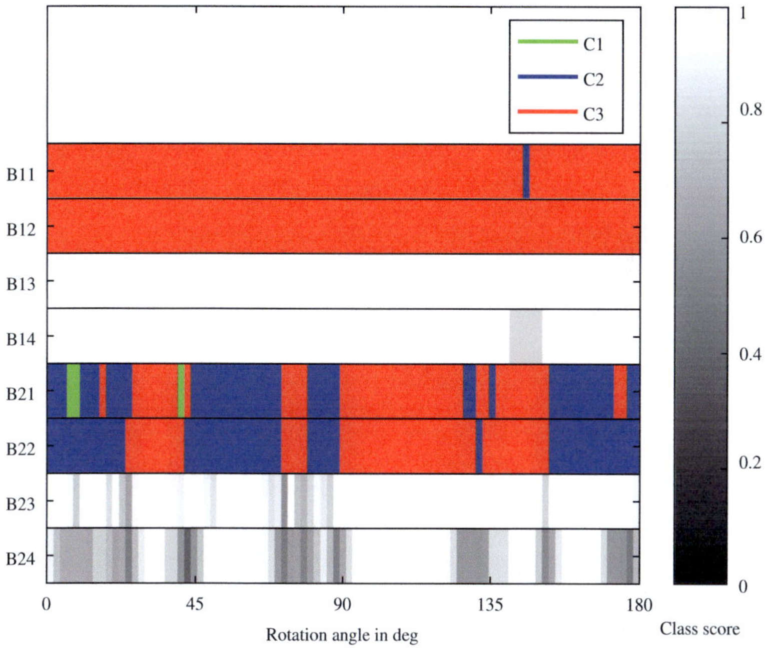

Figure 6.6: Classification decision and class score results of C3O2 from Table 6.1 over rotation angle. Results are presented as barplots for the test datasets at distances of $2.2\,\mathrm{m}$ (labels B1*) and $5.2\,\mathrm{m}$ (labels B2*). From top to bottom: B*1 denotes hard class decision on each rotation angle step, B*2 denotes majority vote of soft class scores for 5 consecutive rotation angle steps, B*3 represents hard class score and B*4 is the soft class score majority vote on 5 consecutive rotation angle steps.

An exemplary RGB polar plot of the polarimetric test datasets is presented for target C3O2 at a distance of $5.1\,\mathrm{m}$ over all measured aspect angles in Fig. 6.5. The polar illustration shows the rotated object as if the radar viewpoint is moving around the target. According to Table 6.1, at approx. $0°$ rotation, a x-pol $45°$-rotated dihedral can be seen. Moving further on to around $90°$, a dihedral rotated by $90°$ comes into view, as well as a box. At around $135°$ to $180°$, another dihedral rotated by $90°$ can be observed.
In Fig. 6.6, barplots of the classification decision and class score results of target C3O2 at distances of $2.2\,\mathrm{m}$ and $5.2\,\mathrm{m}$ are presented. The color format

(a)

I \ O	C1	C2	C3
C1	95	0	5
C2	4	96	0
C3	0	0	100

(b)

I \ O	C1	C2	C3
C1	96	4	0
C2	5	95	0
C3	0	13	87

(c)

I \ O	C1	C2	C3
C1	100	0	0
C2	2	98	0
C3	0	24	76

(d)

I \ O	C1	C2	C3
C1	96	4	0
C2	12	84	4
C3	1	49	51

Table 6.2: Test confusion matrices of majority votes on 5 consecutive rotation angle steps in % over input/output (I/O). 91 rotation angle step measurements per sub-target in a distance of: 2.2 m (a), 3.2 m (b), 4.2 m (c) and 5.2 m (d) are classified. Overall accuracies: 96.9 % (a), 92.3 % (b), 91.2 % (c) and 76.6 % (d).

of the decision bars is assigned in the legend and the colormap of the class score bars is given by the colorbar of the plot. The description of the bars is explained in the figure caption. It is noted that the test measurement at 2.2 m is classified correctly to C3 in bars B11-B14 of Fig. 6.6 for almost all aspect angles. In comparison to that, the classification of test data at 5.2 m in bars B21-B24 is misclassified as C2 in larger piecewise regions of consecutive aspect angles. The bars also show a low class score in many cases of wrong decisions, both for single and majority votes. Reasons for the misclassification are most presumably the similarity of C2 and C3, the above listed physical effects of the radar channel and the limited cells of DOA spectrum. Furthermore, specific learned features of the targets can lead to a one-way misclassification, e.g. C3 is misclassified as C2, but not vice versa. In the following Table 6.2, confusion matrices containing majority vote decisions of the classification results in varying distances are presented. It is noted that with growing distance to the radar, majority vote based classification results degrade significantly. Reasons are probably found in range-dependent physical characteristics of the radar channel. The CNN is only trained with short-distance measurement data in 2.2 m and is not able to learn the physical changes of the radar measurements with growing distance. Nevertheless, as can be seen in Fig. 6.4, the polarimetric information (color information) remains very similar over distance.

6.6 Approaches for Classification Performance Improvement

To overcome the performance degradation of the CNN's classification accuracy on measurement data from untrained distances, three approaches of data normalization, including the global one that has been used for the foregoing results, are discussed. Recalling the effect of changing color intensities, object C3O1 measured in ascending distances, is visible representatively in Fig. 6.7, where the images are normalized to the max. of the Pauli parameters among them. As can also be observed from Fig. 6.4, the effect is barely

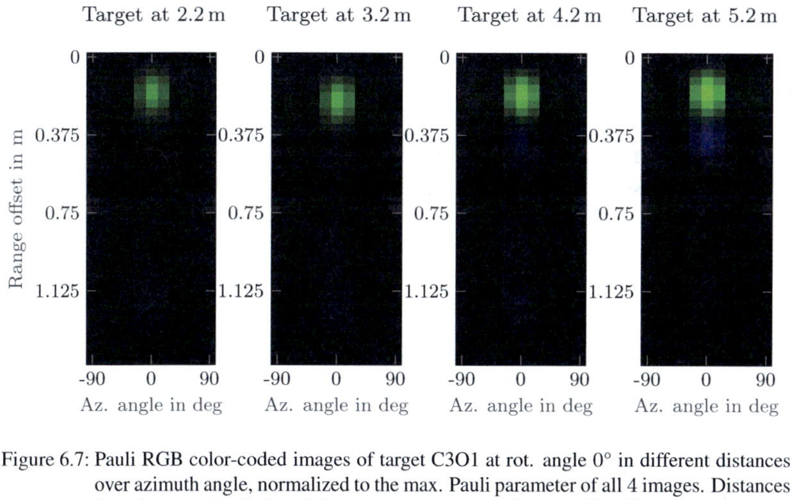

Figure 6.7: Pauli RGB color-coded images of target C3O1 at rot. angle 0° in different distances over azimuth angle, normalized to the max. Pauli parameter of all 4 images. Distances from left to right: 2.2 m, 3.2 m, 4.2 m and 5.2 m. Source: [Sag17].

visible. Therefore, a plot of the normalized maximal color intensity (green color, Pauli c component) from all images in Fig. 6.7 over distance is given in Fig. 6.8. The intensities increase with ascending distance - this is probably the cause of the calibration method that compensates the theoretical free space $1/r^4$ dependence (cf. Section 6.5). It seems that the real radar channel has a deviating characteristic, originating presumably in near-field effects

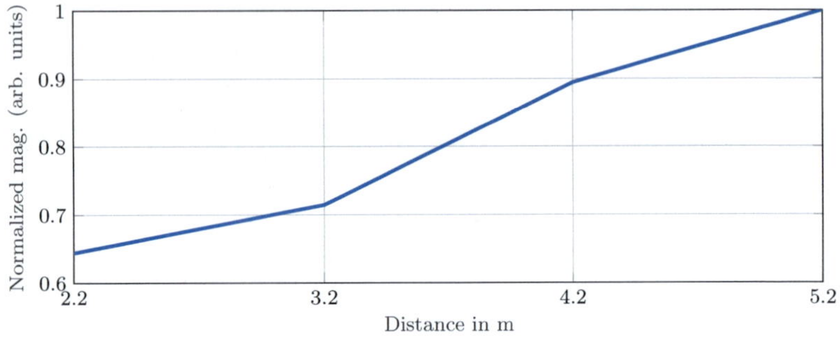

Figure 6.8: Normalized maximal color intensities from the radar images in Fig. 6.7 of target C3O1 over distance. Source: [Sag17].

which decrease the actual RCS in close distances of targets that are very large compared to the wavelength [Kno12].

6.6.1 Normalization Techniques

In order to counteract these varying RCS values of the same target, two other normalization methods are tested. Naturally, the classification accuracies on test data for these approaches are compared with the one using globally normalized data. Exemplarily, Pauli decomposed and globally normalized range-azimuth data of target C3O1 is presented in Fig. 6.9 for three different aspect angles.

For the first approach, a local normalization of RCS values for each radar image snippet is performed. The same dataset as in Fig. 6.9 is presented in Fig. 6.10, where the first approach was applied. It can be seen that in comparison, the colors come out more clearly for the human eye. With this approach, only the information about the maximal value of the highest Pauli RCS per image is discarded. Magnitude gradients and the spatial polarimetric RCS distribution are preserved.

A more rigorous approach can be performed by normalizing each pixel individually, to the highest intensity of the three RGB colors, respectively. Thus, RCS magnitude information contained in the radar image snippets is discarded completely. This approach is performed on the same dataset again, resulting in Fig. 6.11. Now, one RGB color component of each pixel is always one, resulting in very color saturated images. Also, abrupt changes to black color,

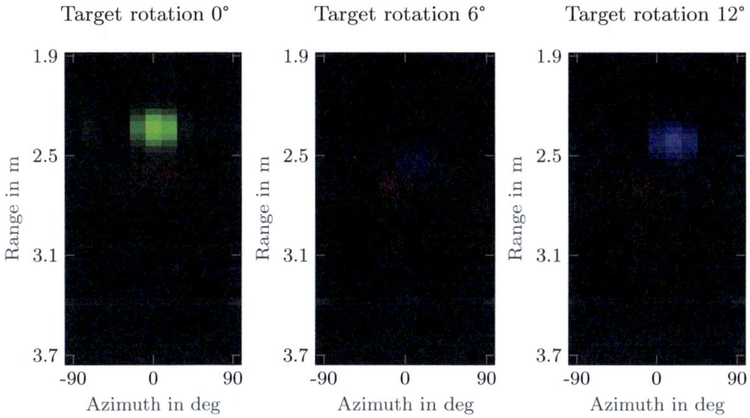

Figure 6.9: Pauli RGB color-coded images of target C3O1 at different aspect angles with global normalization according to the max. of the Pauli parameters in the entire dataset. Rotation angles from left to right: 0°, 6° and 12°. Source: [Sag17].

originating from the introduced $-40\,\mathrm{dB}$ threshold value, are clearly noticeable.

6.6.2 Comparison of Normalization Methods and Usage of an Additional Training Set

In a next step, the CNN's classification performance is evaluated using test data that was preprocessed with each of the three introduced normalization techniques. Also, an additional training set is used, measured at a distance of $5.2\,\mathrm{m}$ with fine rotation angle steps of 0.1° and then preprocessed with global normalization. All trainings and tests are conducted this time using only one target per class, namely C1O1, C2O1, C3O1 and target C3O2' which is basically target C3O2, but mounted in an upright position on the wooden frame. The test set classification performance is presented in form of the resulting test accuracies in Fig. 6.12. With a mean test accuracy of approx. $83\,\%$ to $84\,\%$, both normalization methods result in slightly poorer classification performances than the globally normalized dataset with approx. $85\,\%$ accuracy. Therefore, the normalization methods can be omitted. However, an appreciable improvement of accuracy is reached by conducting the training with the

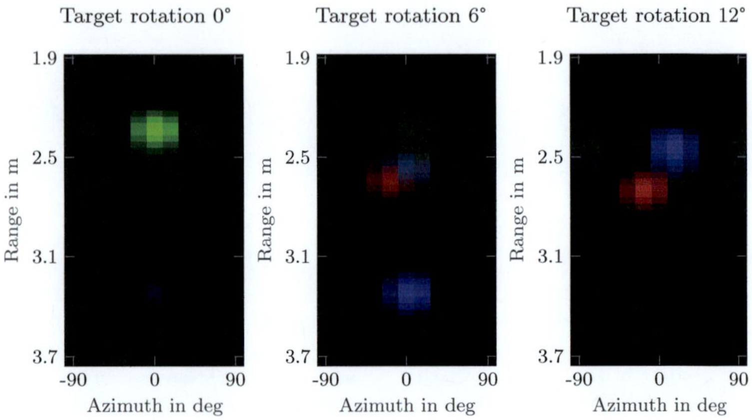

Figure 6.10: Pauli RGB color-coded images of target C3O1 at different aspect angles with local magnitude normalization per image. Rotation angles from left to right: 0°, 6° and 12°. Source: [Sag17].

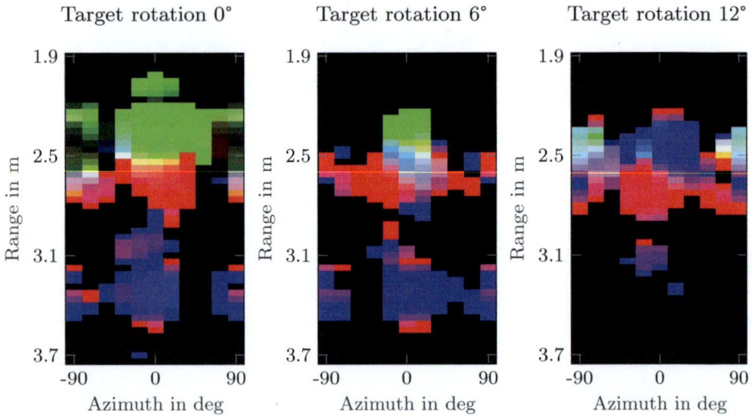

Figure 6.11: Pauli RGB color-coded images of target C3O1 at different aspect angles with magnitude normalization per pixel. Rotation angles from left to right: 0°, 6° and 12°. Source: [Sag17].

additional dataset at 5.2 m distance. Even on the test datasets in intermediate distances, an improvement in accuracy is noticeable. With an overall accu-

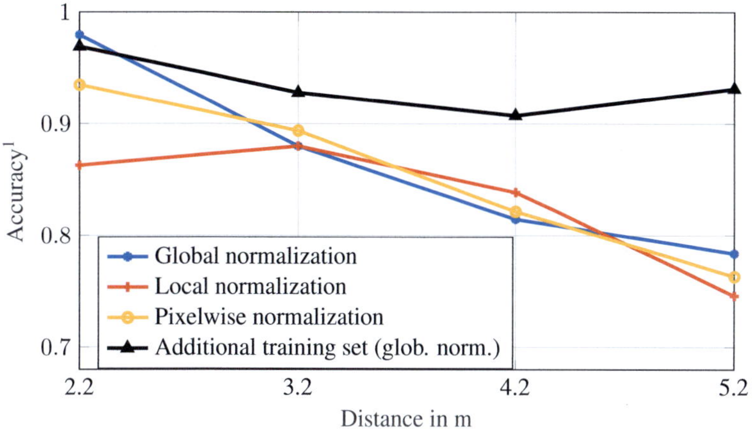

Figure 6.12: Test set classification accuracies for globally normalized (with and without additional training set at 5.2 m distance), locally normalized, and pixelwise normalized datasets. Source: [Sag17].

racy of approx. 93 %, the additional training dataset proves to be necessary in order to reach higher classification performance of the CNN. This heuristic proof is in accordance to the statements in literature (e.g. in [GBC16]) that large training datasets are necessary to reach appropriate generalization and classification accuracies on test datasets, depending also on the specific classification task and the depth of the CNN architecture.

[1] As stated in [GBC16]: "Accuracy is just the proportion of examples for which the model produces the correct output."

7 Classification of Stationary Targets in Vehicular-Based Measurements

Motivated by the promising results of the foregoing chapters, the gained insights are combined to investigate the application of deep learning algorithms concerning more realistic outdoor measurements. For that purpose, the fully polarimetric radar sensor from Chapter 3 with the antenna from Section 3.3 that was calibrated by the method presented in Section 4.1 is mounted in front of a measurement vehicle. The vehicle is also equipped with a differential global positioning system (DGPS)-based localization tracking device. Furthermore, a static measurement scenario with 7 typical traffic relevant targets is set up on a test track. This set of stationary targets comprises: a car, standing and lying motorbikes and bicycles, a stop sign and a construction barrier. While setting up the measurement scenario, care is taken concerning the distances between the targets, so that possible target peak overlaps in the radar spectra are avoided. The utilized FMCW modulation parameters for the radar sensor are listed in Appendix A.2. One resulting limiting factor of these settings and the number of Tx channels operated in TDM mode is a low maximal unambiguous velocity of approx. $3.6\,\mathrm{m\,s^{-1}} \approx 13\,\mathrm{km\,h^{-1}}$. This maximal speed is not exceeded during the measurements. Many varying routes are driven with the measurement vehicle between and around the stationary targets and more than two hours of polarimetric measurement data was recorded, with an update-rate of approx. 20 frames per second.

Subsequent to preprocessing the data per frame and building a ground truth (GT) that comprises measurement vehicle and labeled target map merged into a common geographic coordinate system (GCS), a region of interest (ROI) extractor is used to cut out the parts of the radar spectra where the specific targets are expected respective to the GT.

In a next step, the dataset totaling 130 328 cutouts is analyzed statistically concerning polarimetric target RCS distribution and by using the dimensionality

reduction algorithm t-distributed stochastic neighbor embedding (t-SNE) from [MH08].
Finally, classification of the targets is performed on the spectra cutouts or "snippets". Two classifier algorithms, k-nearest neighbors (kNNs) and CNNs are applied in this chapter. Details about these classifiers can be found in [Fuk90] and [GBC16], respectively.

7.1 Signal Processing Flow Chart

After the measurements were performed, the data is used for classification of the targets. This process can be coarsely divided into the six tasks illustrated in Fig. 7.1. The measurement setup and process itself are described in the following Section 7.2. Two principal branches lead to the final stage of classification.

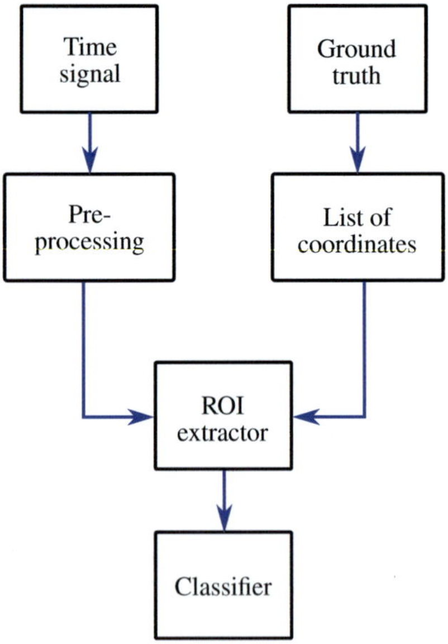

Figure 7.1: Flow chart of classification process - from preprocessing over ground truth to classification.

In the left branch, the time signals of the collected measurement data are preprocessed per frame. This is done in a similar way as in the description in Section 6.4. Furthermore, two equivalent virtual arrays (VAs), one being fully polarimetric and the other one being single VV-polarized, are built in order to compare fully polarimetric and single-polarized performance in a fair way concerning the number of hardware antenna channels. The details of the applied data preprocessing are given in Section 7.3.
Following the right branch of the flow chart, a GT of the measurement site is built, on which basis a list of coordinates, according to the respective radar frame and the targets that are overlapping with the radar beam at the instant of the measurement, is created. Additionally, each target is labeled in this step. Details of this process are described in Section 7.4.
Merging the list of coordinates and the preprocessed radar data, a ROI extractor was implemented. This extractor makes use of the list of coordinates and cuts out the specific part of the radar spectra, where targets are expected according to the GT. The working principle of the ROI extractor is explained in Section 7.5.
Finally, in Section 7.7, CNN and kNN classifiers are trained on these extracted radar image "snippets" and the classification accuracy is evaluated.

7.2 Test Track Scenario and Measurement Vehicle Setup

For the measurement setup, a part of a closed test track was set up with the 7 stationary targets. Aerial photos of these targets can be seen in Fig. 7.2. Some of the targets were encumbered with sandbags in order to stabilize them against heavy winds on the test track. The targets were placed far apart from each other, the minimal distance between them amounted to approx. 10 m, assuring enough free space was provided to move around with the measurement vehicle and at the same time avoiding overlapping target reflexes in the radar spectra.
The ego vehicle was equipped with the radar sensor in the front (mounted at bumper height of about 30 cm) and a localization device with an internal gyroscope in the back. The virtual position of this device was shifted internally to the front of the vehicle so that it coincides with the origin of the radar beam at the sensor. Therefore, the DGPS device provides the origin of the radar beam in longitude and latitude values. Additionally, it provides the ego

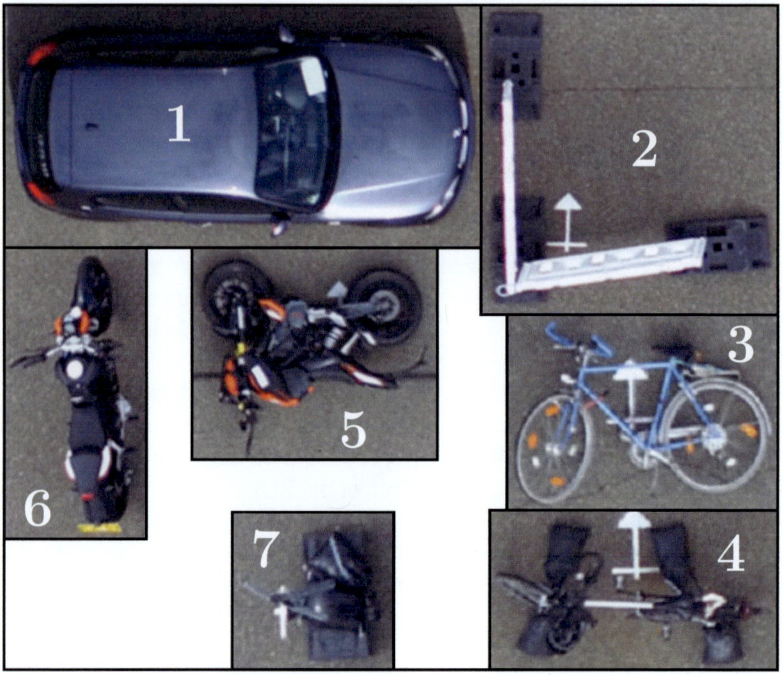

Figure 7.2: Aerial views with labels of 7 stationary targets used for classification. From 1 to 7: Car, construction barrier, lying bicycle, standing bicycle, lying motorbike, standing motorbike, and stop sign.

vehicle's instantaneous velocity and tilt angle φ_{tilt}, which is defined by the angle between the vehicle's central axis and the longitudinal axis (east-west axis), starting also from this axis (cf. Fig. 7.6b). The DGPS device generates approx. 100 measurements per second.

Considering the FMCW modulation and MIMO parameters given in Appendix A.2, a maximal unambiguous range and velocity of approx. $41.7\,\mathrm{m}$ and $\pm 3.6\,\mathrm{m\,s^{-1}} \approx \pm 12.9\,\mathrm{km\,h^{-1}}$ can be calculated using the formulas from e.g. [LB90]. Especially the velocity constraint means that the ego vehicle cannot be moved too fast through the measuring site. Furthermore, the update-rate of the radar measurements/frames amounts to approx. $20\,\mathrm{s^{-1}}$. Finally, the ego vehicle moves in varying trajectories through the setup, trying to catch as many aspect angles of the stationary targets as possible. More than two hours of measurement data is collected in this way.

7.3 Data Preprocessing

The raw time domain data from the radar sensor is first processed by applying a Kaiser window with 70 dB of side lobe suppression (SLS) and a 2D FFT which gives a range-velocity map. By applying a zero-padding factor (ZPF) of two, 512 range bins and 128 velocity bins are available. Then, the FFT-processed data is calibrated by the method described in Section 4.1. After calibration, a motion compensation from [Sch+12] is applied on the data in order to compensate for the movement of the ego vehicle in combination with the TDM operation. Subsequent to these steps, the Pauli decomposition is used to decompose the polarimetric measurement data into co-pol odd bounce (Pauli a), even bounce (Pauli b) and x-pol (Pauli c) parts, as introduced in Section 2.6.1. In order to attain azimuthal information about the targets, a further Kaiser window with 25 dB SLS is applied and a FFT is performed over the spatial dimension of the quasi-ULA antenna elements for each velocity bin, for each polarization combination, and also each Pauli component separately, in a similar way as in Section 5.4.2. Here though, all VA position (channel combination) rows from the redesigned antenna in Fig. 3.8 can be utilized. The two gaps in the arrays at the edges are filled with zeros. The DOA estimation is performed using the method described in Section 2.3.2. Thus, the range-azimuth angle spectrum can be evaluated for each velocity bin and classically for each polarization combination or also for each Pauli component independently. Due to the limited number of array elements, a higher ZPF is applied to the FFT, so that 256 azimuthal angle bins are available.

7.3.1 Fully Polarimetric and Single-Polarized VAs with Equal Channel Count

In order to fairly compare single-polarized and fully polarimetric performance, an equal antenna channel count is needed. Only in this way, the hardware costs of both arrays are comparable. Since the motivation in this thesis is to fundamentally investigate fully polarimetric radars for automotive applications, a full performance system (i.e. where all virtual channels are overlapping) was built in Section 3.3. However, the drawback of this approach, where e.g. all co-

pol channel combinations are overlapping, is that the channel count to reach the same resolution in azimuthal direction as with a single-polarized radar, is much higher. This becomes clearer, if Fig. 3.8 is viewed respective to a single-polarized combination, e.g. VV. Thus, for a fair comparison, VAs with an equal antenna channel count are needed. In Fig. 7.3, one possibility of two comparable 16 channel VAs is illustrated. As can be observed, the fully polarimetric array in Fig. 7.3a comprises 2 orthogonal linearly polarized Tx antennas and respectively 4 Rx antennas for each orthogonal polarization direction. The resulting VA consists of 16 channels, where only 4 channels with a spacing of one wavelength λ can be used for azimuthal DOA estimation. As in Fig. 3.8, the x-pol channels are separated in elevation direction with an arbitrary distance b. Thus, the available azimuthal aperture of this VA amounts to only 3λ, which at 77 GHz is only approx. 12 mm.
In comparison to that, the single-polarized array in Fig. 7.3b comprises also 2 Tx antennas and 8 Rx antennas, which are all vertically polarized and azimuthally aligned, though. The resulting azimuthal aperture of the corresponding VA amounts to 8.5λ, which at 77 GHz is approx. 33 mm. Thus, the aperture size is approx. 2.8 times larger than in the fully polarimetric case, resulting in a much higher azimuthal resolution.
The motivation of the presented approach is firstly that the introduced equivalent VAs can be simply formed from the existing VA in Fig. 3.8. Secondly, this allows a first and fair comparison of the real value of the polarimetric information in comparison to spatial information respective to azimuthal resolution. In addition, the data preprocessing can be performed in the same way as had been described for the fully polarimetric VA in the beginning of Section 7.3.

(a)

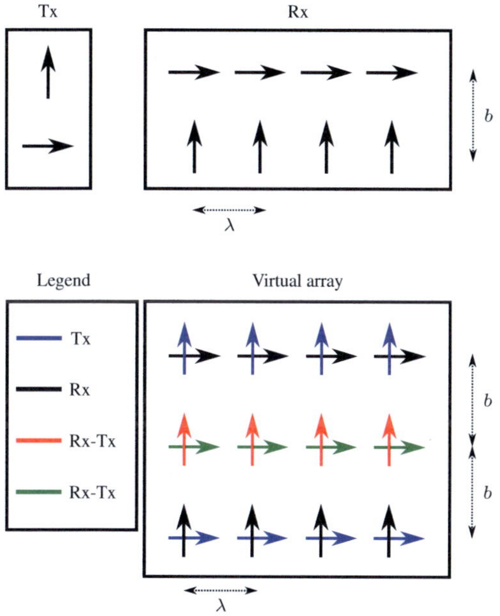

(b)

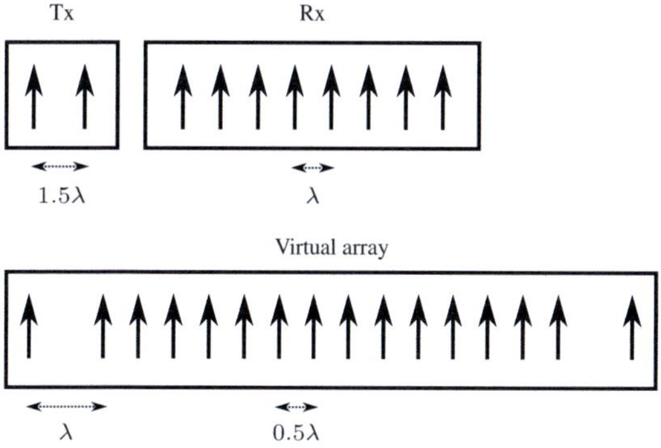

Figure 7.3: Illustration of 16-channel antenna arrays with indicated polarization directions. (a): fully polarimetric version (top image) and corresponding VA (bottom image). (b): VV single-polarized type (top image) and corresponding VA (bottom image).

7.4 Ground Truth Generation

After the stationary targets were positioned, their outlines can be represented by polygons in a local target map. A schematic representation of this local map is given in Fig. 7.4b. The resolution of this gridmap is 3 cm. The ego

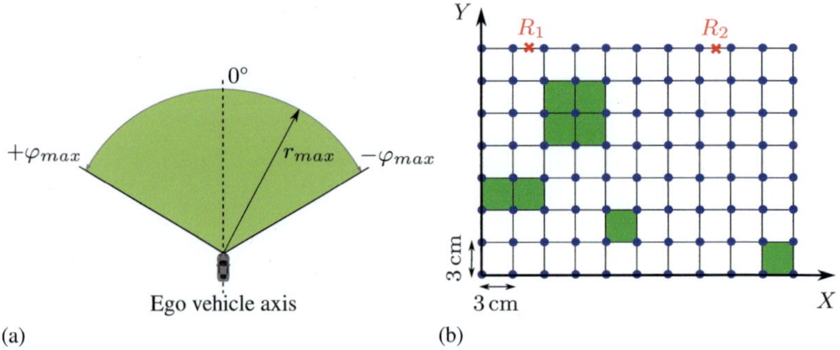

Figure 7.4: Schematic representations of (a): ego vehicle with FOV dimensions: r_{max} = 41.7 m, φ_{max} = 60° and (b): local grid map with a resolution of 3 cm. Green regions illustrate targets in the grid. Checkpoints R_1 and R_2 allow transformation to a different coordinate system. Source: [Rus17].

vehicle is depicted in Fig. 7.4a together with the radar beam in order to show the FOV dimensions. The azimuthal angle φ of the beam is rotated in the mathematically positive direction (counterclockwise (CCW)), starting from $-\varphi_{max}$, and is confined to $\pm\varphi_{max} = \pm 60°$ which corresponds to the 10 dB beamwidth of the sectoral horn antenna elements (cf. Fig. 3.5). The DGPS device of the ego vehicle provides latitude and longitude data in degrees, which correspond to the world geodetic system 1984 (WGS 84) [Kra02]. By using a transverse Mercator map projection, the WGS 84 system can be transformed into a Gauss-Krueger coordinate system [GGG55]. This Gauss-Krueger-map is chosen as the common GCS. In order to generate a GT of the measurement setup, the local target map must be aligned with the ego vehicle map in Fig. 7.5a. Thus, the coordinates of two reference checkpoints R_1 and R_2 situated on a line on the test track are measured very accurately. This allows transformation of the created local target map into the GCS of the ego vehicle by rotation of the local target map around the checkpoint R_1. This procedure is illustrated schematically in Fig. 7.5.

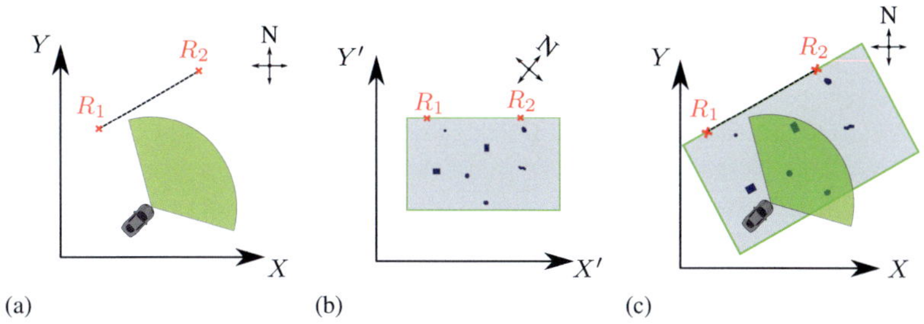

Figure 7.5: Schematic result of the transformation (c) of the target area local map (b) into the ego vehicle map coordinate system (a). Source: [Rus17].

Subsequently, the list of coordinates (cf. Fig. 7.1) can be created. Each target polygon in the GCS is labeled and the central point is defined as the target position. For each measured radar frame, the ego vehicle's position and orientation (given by the tilt angle) together with the radar beam and its origin is assigned to the GCS. An illustration of this method is given in Fig. 7.6a. Since

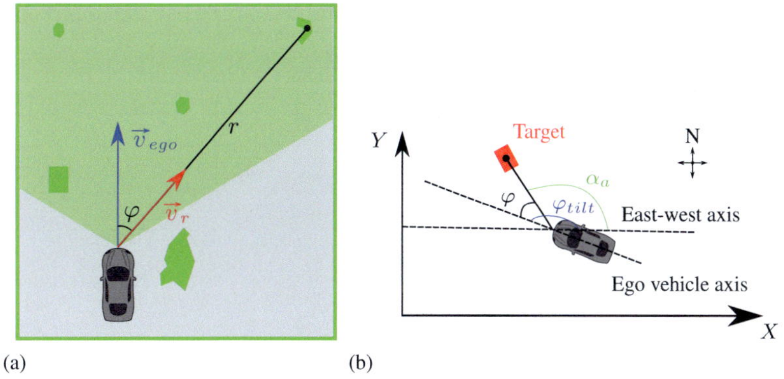

Figure 7.6: (a): Example of target labeling principle with radial distance r from sensor to target center point, azimuthal angle φ, and ego and radial velocity vectors, $\vec{v}_{ego}$ and $\vec{v}_r$, respectively. (b): Schematic calculation principle of the aspect angle α_a. Source: [Rus17].

the DGPS tracking device has a five times higher update-rate, the measurement that is closest to the actual radar frame/measurement in time is chosen for synchronization. For all targets that are overlapping with the radar beam of the specific frame, the radial distance from the origin of the radar beam to the center point of the target can be calculated. Also, the azimuthal angle φ and the radial velocity $v_r = v_{ego}\cos(\varphi)$ are computed. These quantities are also available in the current radar frame's spectra after preprocessing. Consequently, the list of coordinates successfully merges radar beam and targets in the Gauss-Krueger-map. Additionally, the aspect angle α_a of a specific target can be calculated from the GT. For that purpose, Fig. 7.6b is considered. It can be derived that the aspect angle is given by:

$$\alpha_a = \varphi_{tilt} + \varphi \,, \tag{7.1}$$

where the ego vehicle's tilt angle starts from the east-west axis and is rotated in a CCW manner, it is defined as $\varphi_{tilt} \in [-180, 180]$ deg. With its own correct rotation sense and sign (cf. Fig. 7.4a), the azimuthal angle φ of the radar beam is added/subtracted. As a consequence, the aspect angle can also be defined as $\alpha_a \in [-180, 180]$ deg. While facing a target in boresight direction, it starts at zero for a vehicle orientated towards west and becomes $180°$ for an orientation towards east.

All in all, the created list of coordinates comprises: target label, radial distance and velocity, azimuthal angle, and the additional values of ego vehicle velocity and target aspect angle.

7.5 ROI Extraction

With the aid of the list of coordinates from the GT and the preprocessed radar frames, it is now possible to go into the radar range-velocity and range-azimuth spectra and find the position of the labeled targets. In order to apply CNNs as deep learning classifiers with fully connected softmax layers at the outputs, some conditions must be met. One is to use radar spectra that contain only one labeled target at once, because softmax outputs can only determine one probability for each of the targets. Another sufficient but yet convenient condition is that the radar spectra cutouts in the dataset are of the same size in order to make the training process much simpler. A further condition is that the input data which is presented to the CNN is given in form of 2D images, so that the filter kernels can operate in an efficient way. Moreover, with regard to saving computer resources and thus computation time, it is much faster to

train and test a CNN with a dataset of smaller image snippets including only the targets than with large images including a lot of noise and irrelevant information.
Considering all these conditions, it becomes obvious that a ROI extractor is needed that takes the information from the list of coordinates and cuts out only the specific part with the expected target from the radar spectra. The dimensions of the range-velocity-azimuthal angle "cuboids" were chosen according to the maximal distance and resolution of the radar sensor and the largest target in the dataset, the automobile (No. 1 in Fig. 7.2). A schematic cuboid with the chosen dimensions is shown in Fig. 7.7. These dimensions are given as:

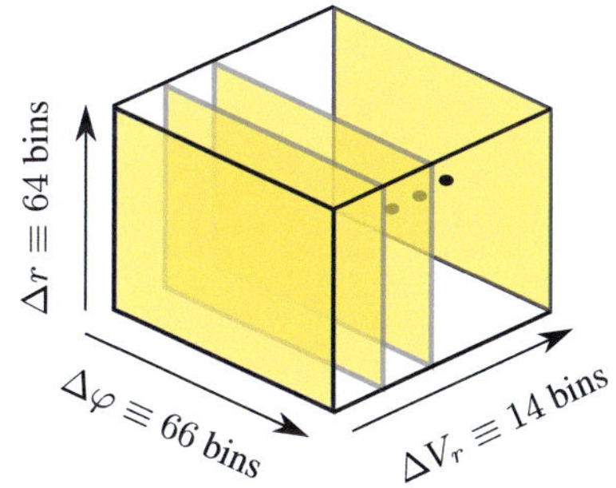

Figure 7.7: Dimensions of a single cuboid created by the ROI extractor. Source: [Rus17].

64 bins in range direction (approx. $5\,\mathrm{m}$), 14 bins in velocity direction (approx. $0.7\,\mathrm{m\,s^{-1}}$), and 66 bins in azimuthal direction (approx. $30°$ around zero, otherwise larger because of the nonlinear arcsin function, cf. Eq. (2.13)). Due to the expected minor relevance of the velocity dimension in the data, originating from the fact that the targets are stationary and only the ego vehicle is moving, not all velocity bins are taken into account for building the dataset. For each of the velocity bins in the extracted cuboid, one range-azimuth image can be chosen. From the available 14 snippets per cuboid, only the one with the maximal RCS is taken, including the two adjacent neighbor snippets on the left and right of it. Thus, the final snippet dataset consists of three range-azimuth snippets per cuboid[1]. For snippets located at the borders of the confined radar beam (cf. Fig. 7.4a), an arbitrary very low padding value of $-150\,\mathrm{dB}$ is set for

[1] If the max. velocity lies at the border of the cuboid less than 3 snippets are taken

the bins lying outside of the beam, in order to maintain the snippet dimensions for the entire dataset.
It is to be noted that by using an ROI extractor of this kind, a lot of correctly labeled snippets that falsely contain only noise and/or clutter are being generated which greatly reduces the quality of the dataset. The reason is that going from the GT, where a target is expected at a specific place, into the radar spectra and cutting out the ROI at that expected place, does not necessarily imply that a reflex of the target was captured in that region at the time the considered frame was measured. In order to make sure that a real reflex from the spectra occurs in the snippets, a peak detector, most reasonably in combination with a constant false alarm rate (CFAR) algorithm, would have to be implemented. However, this approach was not followed in this thesis and is left for future research.

7.6 Dataset Analysis

Once the dataset is completed, it comprises 130 328 radar snippets, nearly equally distributed between the seven targets. In order to find out the highest RCS and the distribution of the RCS values per target, histograms are created. Two exemplary histograms for the motorcycle and stop sign target are presented in Fig. 7.8. The histograms show the number of available snippets over

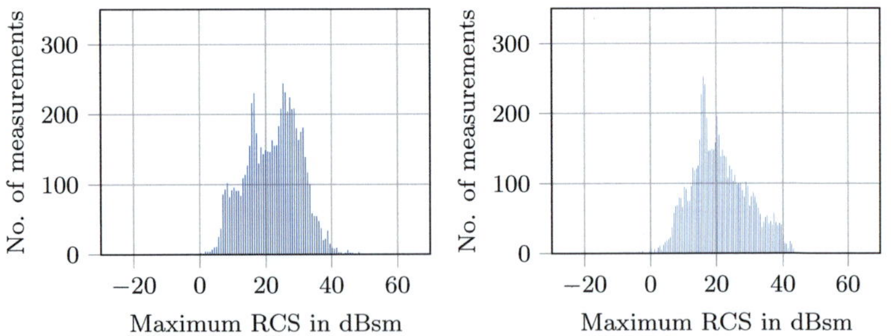

(a) (b)

Figure 7.8: Histograms of local Pauli RCS maxima in fully polarimetric data snippets for standing motorcycle target (a) and stop sign target (b) from Fig. 7.2 (No. 4 and 7, resp.). Source: [Rus17].

the maximum of the Pauli RCS components per snippet (i.e. local maximum), according to Section 2.6.1. Most snippets of the motorbike target in Fig. 7.8a have a higher RCS than 20 dBsm, while most snippets of the stop sign target in Fig. 7.8b show a lower RCS than 20 dBsm. This indicates that the motorbike's reflexes are stronger and lead to a higher signal-to-noise ratio (SNR) in the snippets which in turn could lead to a higher classification accuracy. However, no common probability distribution can be recognized among the histograms. The highest and lowest RCS values of the whole dataset amount to approx. 62 dBsm and the padding value of −150 dBsm. This large dynamic of 212 dB results from the very low padding value. It becomes obvious that a dynamic reduction of the dataset is worth to be investigated.

Recapitulating the results of Section 5.3.1 and Chapter 6, it was found that the targets exhibit a strongly fluctuating polarimetric RCS in dependence of the aspect angle under which they are seen. Therefore, it is expected that the dataset must consist of a lot of varying aspect angles of each target. Exemplarily, the distribution of snippets from the automobile target are plotted over aspect angle in a resolution of 1° in Fig. 7.9. It can be seen that some aspect

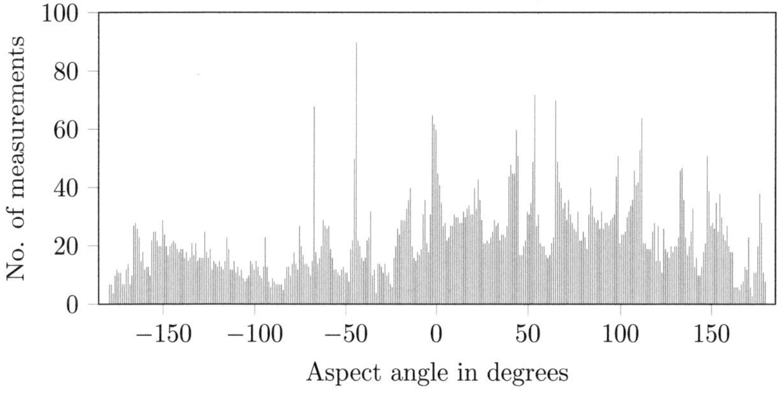

Figure 7.9: Distribution of aspect angles for car target from Fig. 7.2 (No. 1). Source: [Rus17].

angle measurements with that resolution occur less than 10 times in the dataset of the automobile. In retrospect, it was shown already in Chapter 6 that with an aspect angle resolution of even 0.1° the polarimetric RCS fluctuations were still high. Thus, the possibility that the dataset could still be too small in terms of aspect angle information, is given.

7.6.1 Low Dimensional Data Projections

With the t-SNE algorithm from [MH08] it is possible to project a high dimensional feature space into a low dimensional one, mostly for the sake of creating a visually interpretable representation of the dataset. The snippets in the underlying single-polarized dataset (preprocessing and ROI extractor are applied to the dataset of the single-polarized VA from Fig. 7.3b) have dimensions of 64×66 range-azimuth bins/pixels, thus resulting in a high dimensional feature space of 4224 dimensions per snippet. A dimension reduction to 3 dimensions is performed by using the Keras implementation of the t-SNE algorithm. Two exemplary scatter plots, each showing 2000 scatter points/snippets, are presented in Fig. 7.10. In Fig. 7.10a, the low dimensionalized car target scatter

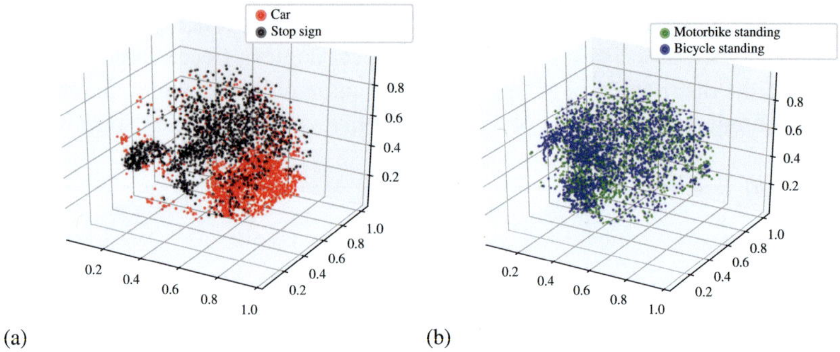

Figure 7.10: Low-dimensional single-polarized data projection results of t-SNE with 2000 snippets. (a): Car vs. stop sign target. (b): Standing motorbike vs. standing bicycle target.

points/snippets are plotted together with the ones of the stop sign target. Interestingly, both of these targets somehow form distinguishable clusters in feature space, meaning that it can be expected that a classification with a simple nonparametric (i.e. the model parameters grow with the training set) classifier like kNN could already yield high accuracies. On the other hand, considering Fig. 7.10b, the motorbike target's scatter points are completely mixed and overlapping with the ones of the bicycle target. In this case, a high classification accuracy by a simple classifier seems not feasible. Of course, this fact results from the similarity of motorbike and bicycle targets, which is reasonable.

In a next step, fully polarimetric data, generated by preprocessing and ROI extraction data of the complete VA from Fig. 3.8, is analyzed. Another approach of dimensionality reduction is used this time. The objective is to isolate the polarimetric information from the spatial one in the dataset, in a similar way as in Fig. 6.3. The magnitudes of the first three Pauli components a, b and c denote the final three dimensions. The abc-voxels of 3260 snippets, mixed from all targets in the dataset, are summed up over their spatial (range-azimuth) dimensions and then normalized to the overall maximum of Pauli components from all the considered snippets. The result is plotted in Fig. 7.11. Each scat-

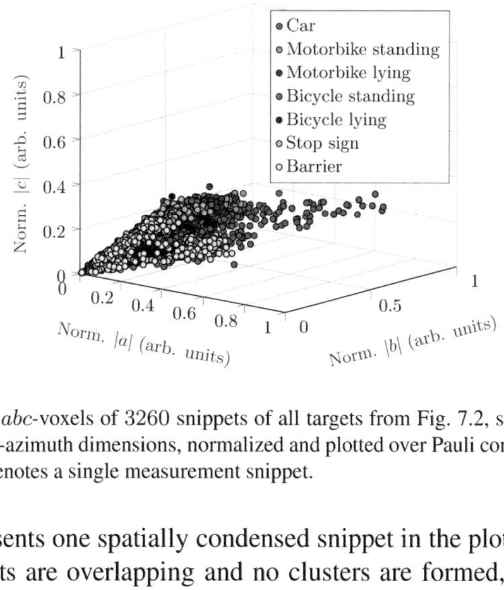

Figure 7.11: Pauli abc-voxels of 3260 snippets of all targets from Fig. 7.2, summed up over the range-azimuth dimensions, normalized and plotted over Pauli component axes. Each dot denotes a single measurement snippet.

ter point represents one spatially condensed snippet in the plot. It can be seen that most points are overlapping and no clusters are formed, indicating that a classification in terms of pure polarimetric information is not possible by using a simple nonparametric classifier. In the following, the importance of polarimetric information versus spatial information for classification purposes is further analyzed.

7.6.2 Evaluation of Fully Polarimetric Data

In this section, selected fully polarimetric radar images of the dataset are presented in the RGB color coded format that was introduced in Section 5.2. In this format, each image is normalized locally (to the highest polarimetric RCS value of the Pauli components in the image) and a certain dynamic in dB is set. In the color scheme of Fig. 5.2 this dynamic amounts to 40 dB. The actually used dynamic is given in the corresponding image captions.

Proof of Concept

An example of a preprocessed fully polarimetric measurement frame, where four trihedral corner reflectors are positioned at the nodes of a 4 m square in a height of approx. 1 m on the test track and are measured from a boresight position and at a distance of approx. 12 m, is given in Fig. 7.12. The result

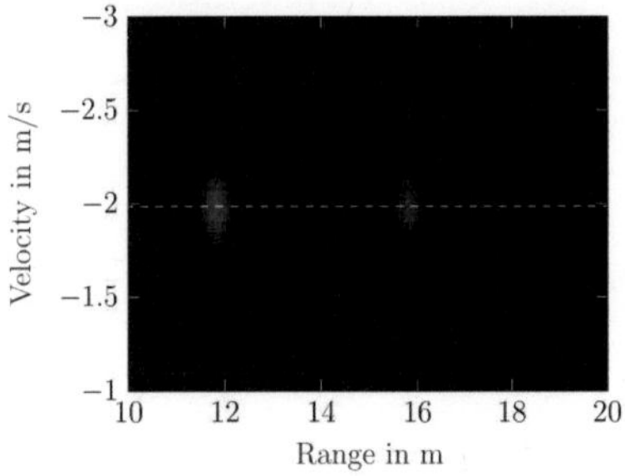

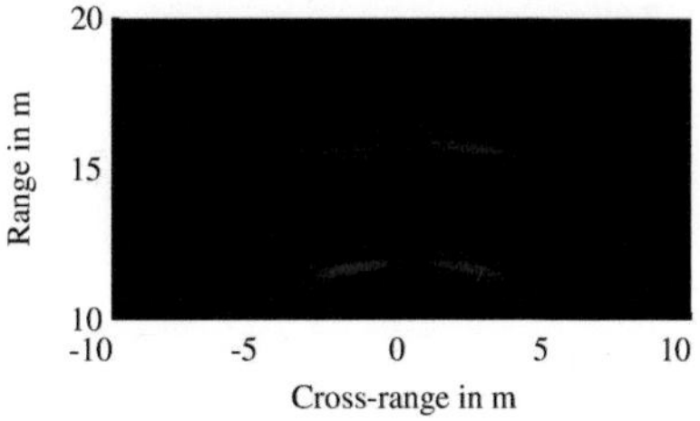

Figure 7.12: RGB color coded Pauli decomposed spectra of four trihedral corner reflectors positioned at the nodes of a 4 m square in a height of approx. 1 m on the test track. Images are locally normalized with a dynamic of 25 dB (cf. Fig. 5.2). (a): range-velocity diagram. (b): range-azimuth diagram at a velocity of $-2\,\mathrm{m\,s^{-1}}$.

was preprocessed by the methods described in Section 7.3 with the 64 channel VA from Fig. 3.8 and is plotted with a dynamic of 25 dB. In Fig. 7.12a, the RGB color coded range-velocity map is shown. Clearly, two blue peaks can be seen at a range of approx. 12 m and 16 m at a velocity of approx. $-2\,\mathrm{m\,s^{-1}}$. The blue color identifies the objects as co-pol odd-bounce objects, where the slightly purple color indicates a small red part. This red co-pol even-bounce part results from the low mounting position of the radar sensor in front of the ego vehicle (approx. 30 cm), whereas the trihedrals are mounted in 1 m height so that a significant part of even-bounce scattering can be observed (cf. Section 5.4.2). In Fig. 7.12b, the corresponding range-azimuth diagram is shown, for the velocity slice at $-2\,\mathrm{m\,s^{-1}}$. Here, the four corners can clearly be separated in their radial distances and angular positions, while the polarimetric

information in form of the RGB colors is preserved. This example proves the preprocessing concepts presented in Section 7.3 and that the radar polarimetry is also working in non-stationary situations.
Another processed measurement frame example is shown in Fig. 7.13. In this

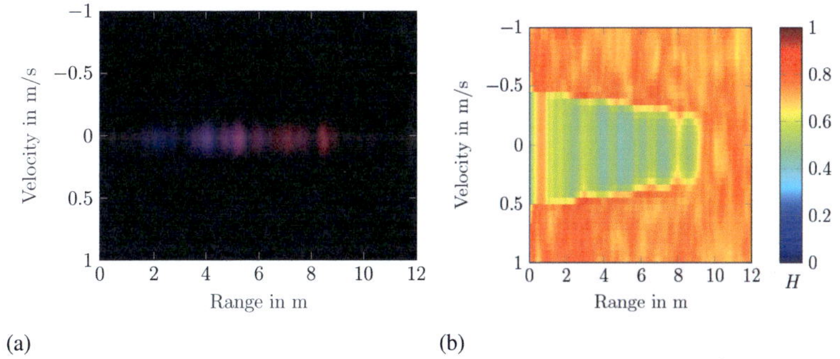

Figure 7.13: (a): RGB color coded Pauli decomposed range-velocity spectrum of a frame measured in front of a vegetated road embankment. Image is locally normalized with a dynamic of 20 dB (cf. Fig. 5.2). (b): entropy of (a), using a 9×9 pixel sliding window averaging.

example, the ego vehicle stopped in front of a vegetated road embankment. The Pauli RGB color coded range-velocity image in Fig. 7.13a is locally normalized and plotted with 20 dB dynamic range. A lot of peaks with varying colors up to approx. 10 m distance already indicate the volume scattering behavior. On this image, a moving average window of size 9×9 pixels is applied. According to Section 2.6.2, this averaging enables a feasible use of the entropy-alpha decomposition. The resulting entropy of the range-velocity image from Fig. 7.13a is shown in Fig. 7.13b. Entropy values higher than 0.5 effectively demonstrate the volume scattering behavior in a quantitative form (cf. Fig. 2.14).

Effects of Varying Target Velocities and Azimuthal Angles

In the following, the dataset is preprocessed with the fully polarimetric 16-channel VA from Fig. 7.3a. In order to demonstrate the difference of the three adjacent velocity "slices" from which the range-azimuth images are built to form the dataset (cf. Section 7.5), Fig. 7.14 is presented. Only subtle changes

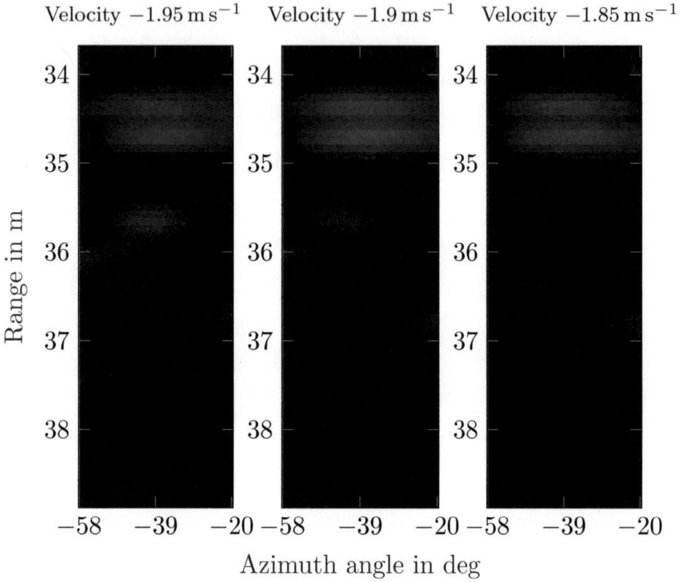

Figure 7.14: RGB color coded Pauli decomposed range-azimuth maps of car target. Locally normalized images with a dynamic of 15 dB Distance: 36.28 m, azimuth angle: $-35.8°$. Velocity from left to right: $-1.95\,\mathrm{m\,s^{-1}}$, $-1.9\,\mathrm{m\,s^{-1}}$, and $-1.85\,\mathrm{m\,s^{-1}}$.

are observed for the example of the car target, thus the method of taking the velocity slice with the max. RCS value and its two adjacent neighbor velocity bins to generate three range-azimuth maps can be considered as a so-called augmentation approach [GBC16]. The small variances in the images help to augment the dataset in a meaningful way. An augmentation of the dataset can help deep learning classifiers to generalize better.

Another important thing to look at are the already mentioned fluctuations of the polarimetric RCS. For that purpose, Fig. 7.15 is shown. With the aid of these exemplary images for the standing motorbike target, it can clearly be demonstrated that a small change of just $0.1°$ in azimuth angle can already lead to large changes in polarimetric information, here represented by the RGB colors. This proves the statement from Section 7.6 that a huge number of aspect angle measurements of the targets must be present in the dataset so that the deep learning classifiers are able to learn the polarimetric features.

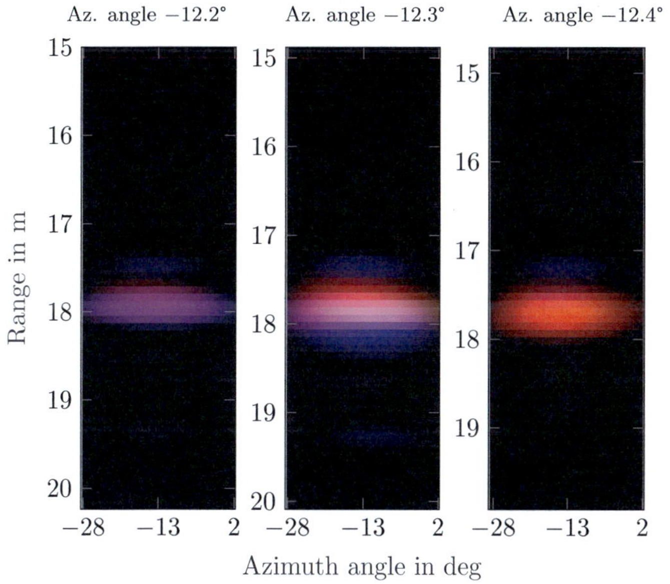

Figure 7.15: RGB color coded Pauli decomposed range-azimuth maps of standing motorbike target. Locally normalized images with a dynamic of 15 dB Distance: 17.32 m, velocity: $-2.74\,\mathrm{m\,s^{-1}}$. Azimuthal angles from left to right: −12.2°, −12.3°, and −12.4°.

On the other hand, less measurements are required for the spatial information, as gets obvious in Fig. 7.15. The changes in the spatial dimension (range and azimuth) are rather subtle in comparison to the polarimetric changes with azimuthal angle.

Normalization Methods

In a further example, different methods of image normalization are compared, similar to Section 6.6.1. The local normalization was already introduced in the beginning of the section, where each snippet is normalized to the highest polarimetric RCS (Pauli parameter) value in the image and a certain dynamic in dB is set. Concerning the global normalization, the highest polarimetric RCS value in the entire dataset is found (cf. Fig. 7.8) and all snippets are normalized to this value. Then, like in the procedure of local normalization, a certain dynamic in dB is set by using a threshold value and setting all values that are below it to the threshold value. In Fig. 7.16, three normalization approaches are presented on the basis of an exemplary measurement snippet of the construction barrier target. In the left image of Fig. 7.16, the global normalization

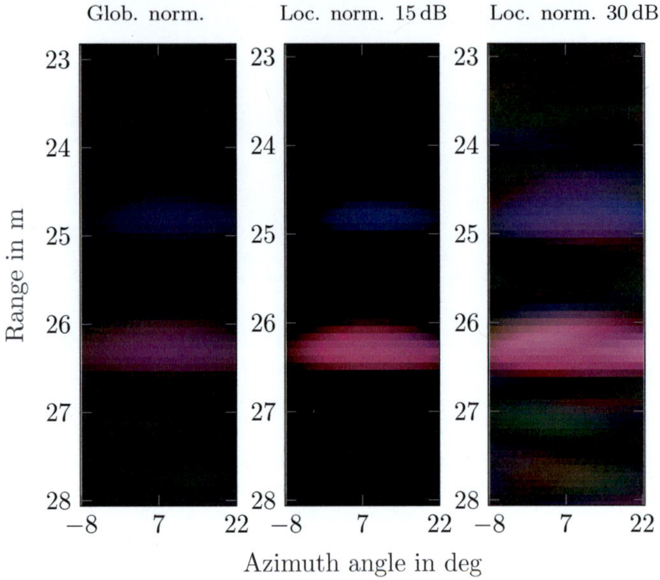

Figure 7.16: RGB color coded Pauli decomposed range-azimuth maps of construction barrier target. Distance: 25.46 m, velocity: $-2.69\,\mathrm{m\,s^{-1}}$, azimuthal angle: 7.3°. Normalization method from left to right: globally normalized with 100 dB dynamic, locally normalized with 15 dB dynamic, and locally normalized with 30 dB dynamic.

with 100 dB dynamic is applied, in the middle image, the local normalization with 15 dB dynamic is applied, and finally, in the right image, the local normalization with 30 dB dynamic is applied. It can be seen that the intensity of the colors for the globally norm. image is much weaker than for the locally norm. ones. This is obvious because the global maximum is approx. 62 dBsm and occurs in the dataset of the car target. As a consequence, all other targets have minor RCS values and the intensity is largely decreased by this normalization method. However, the RCS gradients between all targets are preserved in the entire dataset which could represent a learnable feature for a CNN classifier. In Section 7.7 it is investigated in which way this is beneficial for the classification accuracy. In comparison to that, the local normalization only maintains local RCS gradients per snippet. Furthermore, it can be seen that a doubling of dynamic range from 15 dB to 30 dB results in a more blurry image with more noise, but also more possible target reflexes and polarimetric information can be detected. As these features visually affect the images, they also affect the CNN classifier.

Application of Entropy / Alpha Angle Decomposition

In the last example of this section, a comparison between Pauli decomposed and further averaged and entropy-alpha decomposed images is conducted. Therefore, an example of the car target dataset is chosen and preprocessed with the fully polarimetric VA from Fig. 7.3a. The result is shown in Fig. 7.17. Two predominant peaks can be seen in the Pauli image, the lower one is a clear co-pol even-bounce peak, while the upper one is a mix of co-pol odd and even-bounce. The entropy-alpha decomposition results show a very low entropy in a range of 0 to 0.25 at the peak positions. Furthermore, the alpha angle classifies the lower peak as emerging from a dihedral scatterer, whereas the upper one is clearly identified as coming from a surface scatterer. Finally, the orientation angle θ shows both peaks in an orientation of approx. $0°$ to $10°$. These results are as expected and demonstrate the feasibility of the entropy-alpha decomposition approach.

7.6.3 Comparison of Fully Polarimetric and Single-Polarized Data

After having assessed a few fully polarimetric example images to show some features from this part of the dataset, a comparison of the fully polarimetric 64 and 16 channel VAs (from Fig. 3.8 and Fig. 7.3a) and the VV single-polarized array (from Fig. 7.3b) is conducted. For the single-polarized image visualization, a similar way of locally normalization as for the fully polarimetric images is used. In the same manner, each image is locally normalized to the max. RCS occurring in it, and then a threshold value is applied in order to set a fixed dynamic. The only difference is that there is no RGB color scale, but instead, a single color/grayscale is used. The colorbar of such a grayscale for locally normalized single-polarized images is given in Fig. 7.18. For all the following single-polarized images, this colorbar with the 15 dB dynamic range is utilized. In order to compare the results, an example from the dataset of the automobile target is preprocessed with all three VAs and the images are shown in Fig. 7.19. If the left and middle images are compared, it can be seen that the angular resolution of the single-polarized array with 16 channels in azimuthal direction is the same as the one from the 64 channel fully polarimetric array. This is obvious since the single-polarized VA just comprises the complete VV MIMO block of the central VA channel combinations in Fig. 3.8. Furthermore, there are more peaks occurring in the fully polarimetric image,

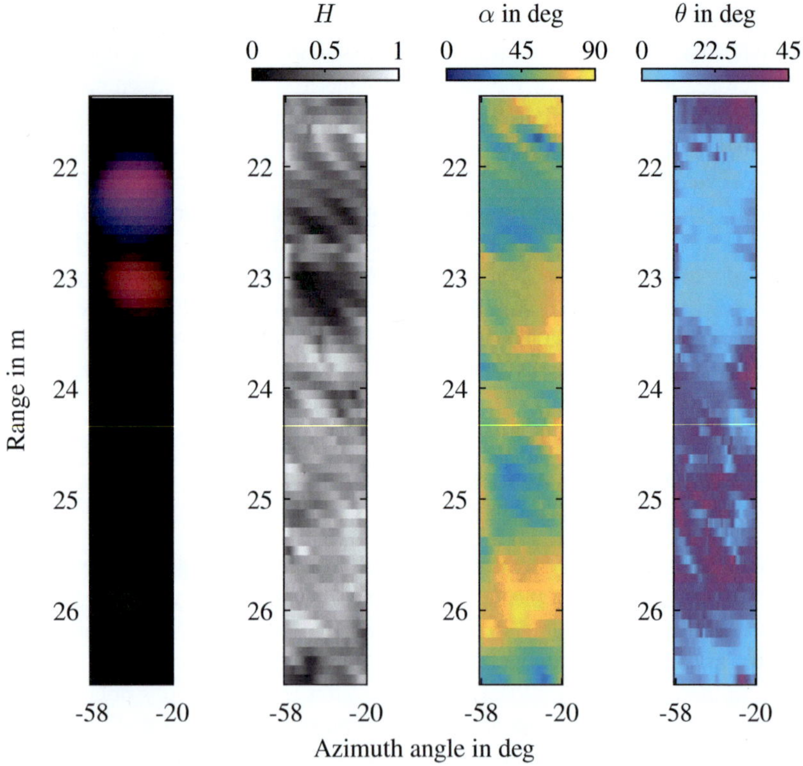

Figure 7.17: From left to right: RGB color coded Pauli decomposed range-azimuth map, locally normalized with 15 dB dynamic. Entropy, alpha angle, and orientation angle of left image, using a 5×5 pixel sliding window averaging. Target: car, distance: $23.02\,\mathrm{m}$, velocity of slice: $-2.07\,\mathrm{m\,s^{-1}}$, azimuthal angle: $-36.7°$.

Figure 7.18: Grayscale colorbar for locally normalized VV single-polarized radar images with a dynamic of 15 dB.

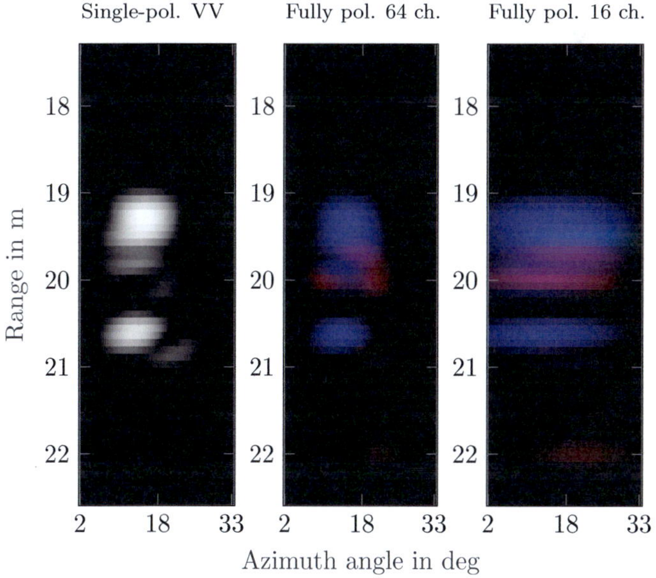

Figure 7.19: VV single-polarized (left) and RGB color coded Pauli decomposed (64 and 16 channel VAs middle and right, resp.) range-azimuth maps of automobile target. Distance: 19.94 m, velocity: $-2.81\,\mathrm{m\,s^{-1}}$, azimuthal angle: 17.3°. Images are locally normalized with a dynamic of 15 dB.

demonstrating the increase in information. In comparison to that, the right image in Fig. 7.19 shows the much smaller angular resolution resulting from the 16 channel VA of Fig. 7.3a. The peaks are much broader and smeared in comparison to the middle image.

A final example is shown in order to demonstrate how a noisy or cluttered image from the dataset looks like. Equally to Fig. 7.19, Fig. 7.20 is created. An example of noisy images from the standing bicycle target is chosen here.

As can be seen, no clear peaks are detectable in the images. A lot of noisy images are found in the datasets of all targets, with an increasing number for targets with a low RCS like e.g. the lying and standing bicycles, confirming the statement of the last paragraph in Section 7.5.

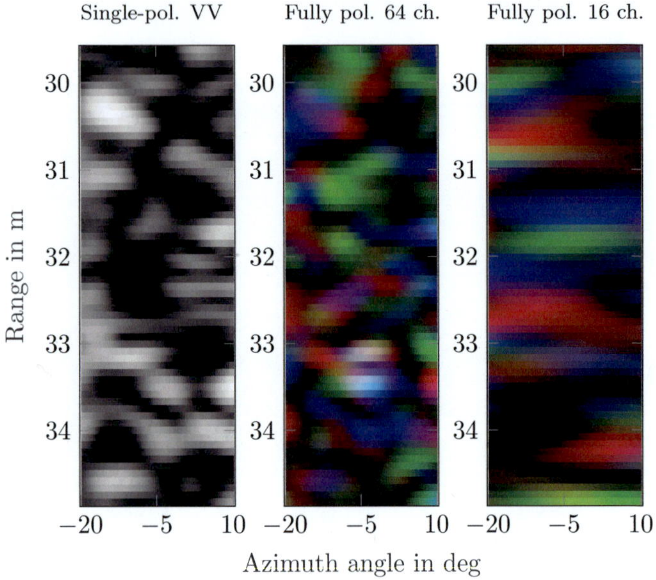

Figure 7.20: VV single-polarized (left) and RGB color coded Pauli decomposed (64 and 16 channel VAs middle and right, resp.) noisy range-azimuth maps of standing bicycle target. Distance: 32.22 m, velocity: $-3.02\,\mathrm{m\,s^{-1}}$, azimuthal angle: $-4.4°$. Images are locally normalized with a dynamic of 15 dB.

7.7 Classification of Targets Using Deep Learning

This section deals with the final block of Fig. 7.1, the classifier. Two classifier algorithms are applied in this section, kNNs and CNNs. The kNN algorithm is used with $k = 1$ neighbor as a nonparametric baseline algorithm for classification, in order to compare the results with the deep learning CNN classifier results. The utilized CNN architecture is depicted in Fig. 7.21. The input

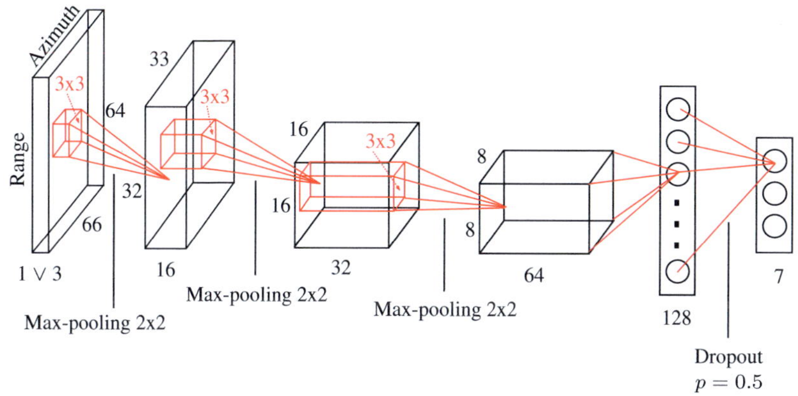

Figure 7.21: Schematic architecture of the used CNN with input and output data dimensions. Input dimension 1 for single-polarized data and 3 for the first 3 Pauli parameters.

data stems from the generated snippet dataset, where the snippets have dimensions of 64×66 pixels (in range-azimuth) and the color depth, which is 1 in case of the 16 channel VV single-polarized VA from Fig. 7.3b and 3 (the first three Pauli parameters) for the fully polarimetric 64 and 16 channel VAs from Fig. 3.8 and Fig. 7.3a, respectively. It is to be noted that the data input is always provided in dB, in combination with the normalization techniques introduced in Section 7.6.2. Three convolutional layers with an increasing number of 3×3 filter kernels (16, 32, and 64) are used in combination with 2×2 max-pooling layers. At the end, a fully connected layer with 128 neurons is followed by another fully connected softmax layer with 7 neurons for the class score outputs. A dropout of 0.5 is used for the purpose of generalization and in order to prevent overfitting. Other applied methods to facilitate regularization are the L^2 norm weight decay and the early stopping method [GBC16]. Like in Section 6.2, the Adam algorithm is used for adjusting the learning rates during the training process, starting from predefined values between 10^{-5} and 10^{-4}, and the cross-entropy loss function is applied to the softmax output layer. A minimum of 500 epochs are trained in order to optimize the model. For all following results, the complete dataset was split into 80 % training data, 15 % validation data and 5 % test data. The difference of validation and test dataset is that the validation data is used during the CNN training process to optimize the model. The test dataset is unused until the final model is chosen according to the early stopping method [GBC16]. A classification of the seven targets by pure chance amounts to an accuracy of approx. 14.3 % (1/7).

7.7.1 Comparison of Global and Local Dataset Normalization

Given these settings, a first investigation concerning the global and local normalization methods with different dynamic ranges is conducted. As dynamic range for the global normalization, the total dynamic of the dataset from global maximum to the low padding value of −150 dB is taken, cf. Section 7.6. Therefore, the datasets are preprocessed with the different normalization methods and the CNNs are optimized in terms of validation set accuracy. The final results after optimization are presented in Table 7.1. There

Norm. type / Array type	Global norm. 212 dB dynamic	Local norm. 60 dB dynamic	Local norm. 30 dB dynamic
Fully polarimetric 64 channels	68 %	68 %	69 %
VV single-polarized 16 channels	58 %	—	57 %
Fully polarimetric 16 channels	58 %	—	58 %

Table 7.1: Comparison of classification accuracies on the validation set for different array type configurations and varying normalization approaches.

is no obvious advantage in using the extremely large dynamic of 212 dB in combination with the global normalization. Instead, a much more comprehensible local normalization leads to equal results, even with a smaller dynamic of 30 dB. This means that the RCS gradients between the targets and frames do not contribute to more information about the targets, as discussed in Section 6.6.2. The local RCS gradients in the images hold enough information for the classifier. The classification accuracy of the kNN algorithm is at least 5 % lower in all cases. Thus, the nonparametric classifier reaches also high classification accuracy results. The reason can be found in the high redundancy of the dataset, originating from the augmentation method of Section 7.6.2 and the high spatial dimension similarity of the snippets in the dataset, cf. Fig. 7.15. The kNN algorithm performs well on spatially similar datasets because it reduces each snippet from the training dataset to a point in feature space and then calculates the distances of the k neighbors to the actually tested snippet from the validation dataset. Thus, the training set in feature space is similar to the t-SNE illustrations in Section 7.6.1. Nevertheless, the CNN algorithm still performs better in this experiment. Also, the classification is computationally more efficient, once the final CNN model has been trained [GBC16]. The highest baseline classification accuracy is reached with the 64 channel fully polarimetric VA, amounting to approx. 70 %. In comparison to that, the other two VAs perform with an approx. 10 % lower classification accuracy on the seven targets.

Promisingly, the 16 channel fully polarimetric VA shows the same classification accuracy results as the 16 channel single-polarized VA, although it has an almost three times larger azimuthal aperture. These important results signify that the CNN is able to learn some of the polarimetric features, and that an approx. three times larger azimuthal resolution can be fully compensated with the polarimetric information.

Furthermore, the CNN is tested with preprocessed four dimensional input data for the fully polarimetric arrays, meaning that all parts of the (a, b, c, d) Pauli parameter's RCS are taken into account, instead of only using the first three ones. As expected, the results do not yield higher classification accuracies since the Pauli d parameter comprises mainly insignificant information for a quasi-monostatic system [LP09].

Also, the entropy-alpha decomposition with a 5×5 sliding average window is appplied to the fully polarimetric datasets. The classification accuracy results on the validation datasets dropped about almost 20 % after the CNN trainings with the entropy, alpha angle, and the orientation angle as data input dimensions (cf. Fig. 7.17) had been conducted. Thus, the entropy-alpha decomposition is not considered beneficially for the preprocessing of the dataset in terms of classification accuracy of the seven considered targets. This is a good result, as the entropy-alpha decomposition is computationally expensive, mostly due to the eigenvalue/singular value decomposition that needs to be performed for each snippet in the dataset (cf. Section 2.6.2).

Finally, the results indicate that a local normalization of the snippet dataset with a 30 dB dynamic is sufficient to reach good results of the CNN classification accuracy. Another finding is that with the complete dataset, the highest classification accuracy amounts to only about 70 %. The reason for this low accuracy is most presumably found in the high amount of noisy/cluttered snippets in the dataset, which are correctly labeled as the resp. targets, but instead bear no usable information.

7.7.2 Classification Results Using a Filtered Dataset

In order to counteract the bad quality of the dataset, a simple filter algorithm, based on the cell-averaging CFAR algorithm from [SD91], is implemented. The filter calculates the local RCS average of all pixels/voxels in the underlying snippet and compares it with the local maximum. If a certain threshold dynamic between these values is undercut, the snippets are eliminated from the dataset. At the end, the percentage of eliminated to remaining snippets is calculated, according to the threshold value. With this method, the signal-to-clutter-ratio of the remaining snippets in the dataset can be greatly enhanced, which in turn increases the quality of the dataset to a high amount. The best classification accuracy result could be reached for a 50 % filter, i.e. 50 % of the whole dataset is eliminated by the filter, resulting in a reduced dataset of approx. 65 000 snippets. The optimized CNN's training and validation accuracy, and corresponding loss functions are plotted over 500 training epochs in Fig. 7.22. The underlying dataset was preprocessed using the 64 channel fully polarimetric VA from Fig. 3.8 and a local normalization with 30 dB dynamic is applied. As can be seen, 90 % of classification accuracy could be reached on the validation dataset. This is the best result that could be reached on the dataset with the seven targets. Moreover, the classification accuracy of the kNN classifier with $k = 1$ neighbors also increased rapidly and lies only approx. 5 % lower than the one of the CNN. The results of the preprocessed datasets of all three VAs with different filter percentages are presented in Table 7.2. Obviously, the accuracy of the optimized models rises up to a filter percentage of 80 %. On the other hand, if more features are available, e.g. full spatial and polarimetric resolution in case of the fully polarimetric 64 channel VA, the accuracy decreases again slightly. That indicates the turning point for this VA, so that for higher filter values than 80 % the dataset becomes too small for all features to be learned by the CNN. For the fully polarimetric 16 channel VAs, the second highest validation set accuracy of 84 % could be reached.

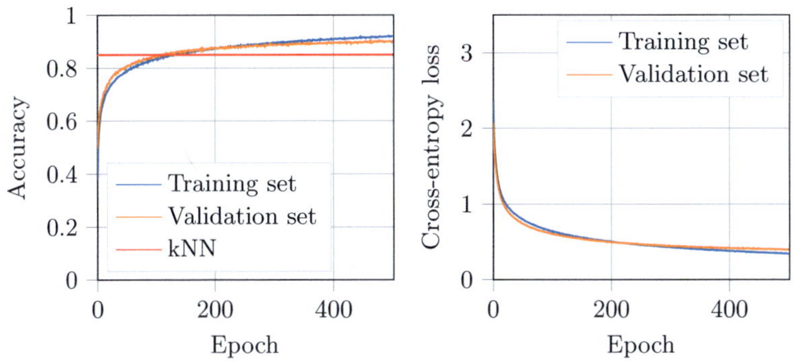

(a) (b)

Figure 7.22: CNN model accuracies (a) and losses (b) for 50 % filtered dataset. Locally normalized data that was preprocessed using the 64 channel fully polarimetric VA from Fig. 3.8, with a dynamic of 30 dB.

Filt. amount / **Array type**	**0 %**	**10 %**	**25 %**	**50 %**	**80 %**
VV single-polarized 16 channels	58 %	61 %	69 %	79 %	82 %
Fully polarimetric 16 channels	60 %	62 %	69 %	78 %	84 %
Fully polarimetric 64 channels	70 %	75 %	79 %	90 %	89 %

Table 7.2: Comparison of classification accuracies on the validation set for different array type configurations and filter amounts. Total dataset consists of 130 328 snippets. Locally normalized data with a dynamic range of 30 dB [Rus17].

At the end, the confusion matrix of the test set accuracies for all targets, resulting from Fig. 7.22, is given in Table 7.3. The average classification accuracy on the test set amounts to approx. 88 % which comes very close to the validation set accuracy. This result proves the good generalization of the utilized CNN model. The best classified target in the dataset is the car, with the highest accuracy result of almost 96 %, closely followed by the construction barrier target with approx. 94 %. These targets are the biggest targets and therefore most likely exhibit also the highest RCS of the dataset.

I \ O	Car	Motorbike standing	Barrier	Motorbike lying	Stop sign	Bicycle	Bicycle lying
Car	95.64	1.33	1.7	0	0.38	0.19	0.76
Motorbike standing	0	83.76	0.74	0.37	4.8	6.64	3.69
Barrier	1.89	0.27	94.34	0.27	1.62	1.35	0.26
Motorbike lying	0	1.46	0.97	78.16	2.43	7.28	9.7
Stop sign	1.24	2.48	0.62	0.31	93.8	0.31	1.24
Bicycle	0.3	4.14	0.3	2.65	0.3	85.21	7.1
Bicycle lying	0.48	3.83	0.96	3.83	1.91	5.74	83.25

Table 7.3: Test set confusion matrix of CNN model results from Fig. 7.22. Classification accuracies for all classes/targets from Fig. 7.2 in % over input/output (I/O).

Additionally, the confusion matrix shows clearly the false positives and negatives. The highest misclassification percentage amounts to approx. 10 % and corresponds to the targets with the weakest RCS in the dataset - the lying motorbike and lying bicycle, respectively. Following, standing and lying motorbikes and bicycles are also misclassified with the highest percentage. These results are reasonable since targets with a low RCS provide snippets with lower SNR values and thus the dataset suffers in quality. Accordingly, less features can be learned by the CNN algorithm which in turn leads to a higher misclassification percentage in the confusion matrix. A way of increasing the classification accuracy of targets with a weak RCS could be to use a radar with higher resolution and/or higher sensitivity. This approach is left for future works, though.

8 Summary and Conclusion

This thesis focused on the investigation of polarimetric radars for automotive applications at 77 GHz. The outcome of an initial literature review exposed that there was little to none published research work on the topic. Motivated by promising results in a growing number of fields such as synthetic aperture radar (SAR) polarimetry, ground penetrating radar (GPR), radar imaging, weather radar, and road surface condition classification, an analysis of radar polarimetry was conducted and the applicability to radars used in automotive vehicles was researched. Since the entire polarimetric scattering matrix can be measured with a fully polarimetric radar, up to five times more information about the observed scene within the field of view (FOV) is obtained. Thus, it could be expected that with this additional information improvements in topics such as target classification and mitigation of multipath effects (such as ghost targets) could be achieved.

First, an experimental frequency-modulated continuous-wave (FMCW) radar sensor was equipped with an orthogonal linearly-polarized waveguide antenna comprising 16 sectoral horn elements with high polarization isolation of more than 30 dB. Using this antenna in combination with a time-division multiplexing (TDM) multiple-input, multiple-output (MIMO) approach, a fully polarimetric and real-time capable FMCW radar sensor operating at 77 GHz was built. Leveraging the MIMO approach, a virtual array (VA) of overlapping receiver (Rx)-transmitter (Tx) channel combinations with orthogonal polarizations was formed. This novel approach enabled the coherent extraction of the important relative polarimetric phase differences in order to measure the full complex scattering matrix (S-matrix), independently of the target position. In contrast to that, only one entry of the 2×2 S-matrix can be measured with a single-polarized radar.

Next, a new radar calibration technique for the sensor was developed. It based on the isolated antenna calibration technique (IACT) from [SUT90] and was generalized in this work to allow calibration of also bistatic and quasi-monostatic MIMO radars operating with antennas of high polarization purity. During calibration, a co- and a cross-polarizing target must be placed one by one in boresight direction at a certain distance to the radar. Subsequent to the calibration, multi-angular measurements in an anechoic chamber were performed on canonical objects to prove the calibration method and to demonstrate the occurring polarimetric effects in millimeter wave (MMW) range. Furthermore, coherent and incoherent decompositions of the S-matrix were successfully applied to the measurement data, which had only been used in SAR polarimetry before.

Then, a rigorous analysis of multipath and direction of arrival (DOA) detection, focusing on MMW radar polarimetry, was conducted. The theoretical analysis revealed that multipaths originating from over-the-road reflections bear minor polarimetric detection potential due to the rather grazing angle of incidence on the road in automotive applications. However, multipaths originating from metallic surface reflections change the polarization state and were thus investigated with a measurement setup in an anechoic chamber. From the measurement results, promising approaches of improving DOA detection in the presence of multipath reception were elaborated.

In a further step, the object classification potential using polarimetric data was investigated. For this purpose, several two-wheeled vehicles, as representations of vulnerable road users, were measured in an anechoic chamber on a rotary table. The vehicles were turned around in small angle steps and measured by the radar for each angle step at a close distance of a few meters. In this way, polarimetric rotation signatures of the vehicles could be generated. By using the coherent Pauli decomposition method, color-coded radar images were created. These results illustrated the unique spatial shapes and color gradients of the different vehicles over aspect angle and confirmed the potential of high target classification capability.

Following this idea, a machine learning approach using a convolutional neural network (CNN) to perform the classification task was applied. For this purpose, several artificial targets that contained varying numbers of canonical objects, were manufactured. Some of these targets purposely had a high similarity in order to complicate the classification task for the CNN. Rotation measurements on a rotary table were performed with these targets, in the same way as with the two-wheeled vehicles. A CNN topology, originally used for the classification of a large publicly available dataset of handwritten digits, was modified to classify Pauli-decomposed spatial range-azimuth data of the rotated targets. It was found out that training data for different target distances is required in order to reach a good generalization of the CNN. With this novel approach, a high classification accuracy of more than 90 % could be achieved.

After evaluation of the previously obtained results, it turned out that it would be advantageous to improve the azimuth resolution of the radar. Thus, a second dual-orthogonally polarized waveguide antenna was designed with double virtual azimuthal aperture size, but the same number of sectoral horn elements as the first one. This antenna was manufactured using an innovative stereolithography apparatus (SLA) 3D printed and copper-coated layer stack technique.
Finally, an experimental FMCW radar was equipped with the improved antenna, and was used for real world outdoor measurements of 7 stationary and traffic relevant targets on a test track. These representative targets included an automobile, standing and lying motorbikes and bicycles, a stop sign and a construction barrier. To take measurements, the polarimetric radar sensor was installed in front of a car. Numerous varying routes were driven between the static targets and more than two hours of measurement data was captured. A ground truth (GT) was built by merging the data of a global positioning system (GPS) navigation device that had been installed in the measurement vehicle, and an aerial map with labeled polygons as target locations. After preprocessing the radar data, a region of interest (ROI) extractor was implemented in order to utilize the positions from the GT to cut out cuboids from the radar's range, velocity and azimuthal spectra. Using those radar image "snippets", CNNs were applied to classify the targets. It could be shown that a fully polarimetric radar is advantageous in terms of target classification accuracy by approx. 10 %, compared to a single-polarized one with the same spatial resolution. After filtering of noisy data, a top classification accuracy of approx. 90 % could be reached with the fully polarimetric approach. Even more important seems the result that a fully polarimetric MIMO radar with the same number of hardware antenna channels, but almost three times smaller azimuthal aperture size compared to its single-polarized counterpart, performed equally in terms of target classification accuracy. This means that for classification purposes, the additional polarimetric information is able to compensate an almost three times smaller azimuth resolution. In this case, a top accuracy of approx. 84 % could be reached. The dataset showed though that the polarimetric information of the targets is a much more subtle feature to be learned than the pure spatial imformation. This means in return that much less training data in the spatial domain is needed to train a well-performing CNN classifier in comparison to the polarimetric domain, since this information is coupled to the polarimetric radar cross section (RCS) which is way more fluctuating with aspect angle and is also less correlated.

An additional insight from the test track measurement results was the proof of the assumption that dense vegetation, e.g. high grass at the side of the road, leads to a high depolarization at 77 GHz which can be detected using the incoherent entropy-alpha decomposition.
This outline presented the main achievements attained in this thesis. Several points could be addressed in future research. At first, an optimization of the calibration procedure can be attempted. Performing the calibration not only in boresight direction, but for a fine grid in azimuthal direction enables the computation of improved DOA spectra using super-resolution methods such as multiple signal classification (MUSIC) or estimation of signal parameters via rotational invariance technique (ESPRIT) [Wen+07; WWS08]. Consequently, further improvement of the antenna towards a higher azimuthal DOA resolution and/or additionally applying channels for elevation DOA estimation can be performed to allow contour recognition.
Considering possible cost optimization approaches, a reduction of the number of required overlapping polarimetric channels can be tackled. One conceivable approach would be to implement a partially polarimetric MIMO radar using circularly polarized transmitters and orthogonal linearly polarized receivers which is known as one of the "compact polarimetry" manifestations [Ran07; Che+16]. This method allows saving half the number of Tx channels, because only one circular polarization (right hand circular polarized (RHCP) or left hand circular polarized (LHCP)) is used.
Another important analysis concerning the implementation of advanced machine learning algorithms like semantic segmentation or region-based convolutional neural networks (R-CNNs) [SLD17; Gir+16] can be conducted, using the collected polarimetric test track measurement data. These advanced algorithms could not only enable the classification of targets but additionally provide their localization in the radar images, up to the labeling of single pixels (resolution cells).
Finally, enhanced test track and also real road traffic measurements could be conducted in order to further investigate the classification potential of road users and other targets using machine learning algorithms. These steps could be further divided into a second measurement campaign where not only the ego vehicle is moving, but also the targets. A third measurement campaign could be arranged in real traffic while driving on the highway and/or in the city. The low maximum unambiguous velocity resulting from the Doppler frequency estimation for TDM MIMO FMCW radars could be increased using methods from [Roo+18]. Eventually, also a fusion of other sensor data like

lidar and video data with the radar data like in [BH10] could be performed, in order to merge all sensor data regarding further improvement of classification accuracy and/or to make the entire system more redundant.
This work presented the first in-depth analysis and fundamental approaches to pave the way for future polarimetric radars in automotive applications. The elaborated results clearly demonstrate that radar polarimetry is realizable with modern MIMO radar systems and brings real benefits with it. Based on this work, the fundamental ideas and approaches can be optimized further on in order to create cost-efficient solutions that bring along the best benefits for the most promising use-cases of future automotive radars.

A Appendix

In this appendix, a brief overview of the fast-chirp frequency-modulated continuous-wave (FMCW) radar principle is given and the main modulation parameters of the measurements taken in this thesis are presented. Furthermore, the basic digital signal processing steps are introduced. For more information and detailed mathematical description of the FMCW principle, the interested reader is referred to the respective literature: [Bar73; LB90; Sko02]. For more details on multiple-input, multiple-output (MIMO) radars and digital signal processing, the interested reader is referred to [LS08; Ste00].

A.1 Fast Chirp FMCW Radar Principle and Digital Signal Processing

In Fig. A.1, an overview of the variety of radar types is given in a tree chart. In this thesis, from the large variety of radar types, only quasi-monostatic FMCW radars are utilized. In order to operate the radar in MIMO mode, the transmitter (Tx) antennas of the sensor are switched in time-division multiplexing (TDM) mode. The ramps being used are fast ramps, thus the radar can be considered as a fast-chirp MIMO radar [TRF15]. In the top image of Fig. A.2, the fast chirp principle is schematically demonstrated for one frame, which consists of multiple chirps. The transmitted frequency chirps are only up-chirps, which get reflected by targets in the radar's field of view (FOV). The received chirps are Doppler and time shifted, according to the corresponding radial velocities and distances of the targets to the radar, respectively. In the example of Fig. A.2, the Doppler shift is positive, meaning that the target relatively moves towards the radar, leading to an increase in frequency. Also, the received ramps are time shifted, meaning that the target is located at a certain radial distance to the radar, being proportional to the time of flight (TOF) from radar to target and back. Per receiver (Rx) channel, the received signals are mixed down, low pass filtered and then sampled with an appropriate sampling rate, fulfilling the Nyquist-Shannon theorem. After being digitized

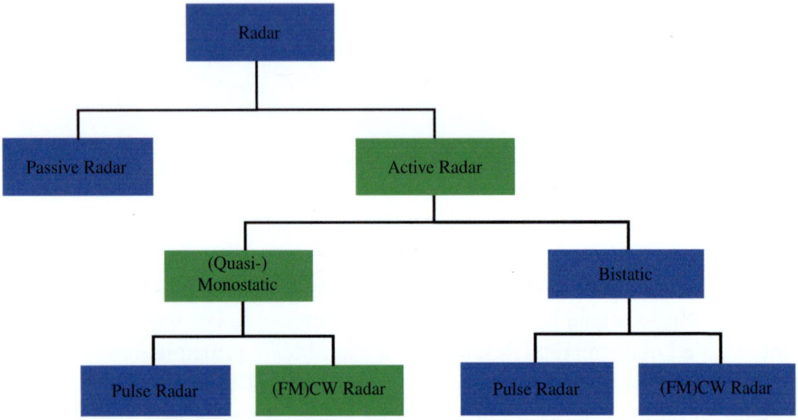

Figure A.1: Tree diagram with an overview of basic radar principles.

by analog-to-digital converters (ADCs), the signals can be digitally processed. The Doppler shift is proportional to the relative radial velocity of the target to the radar and the TOF leads to a beat frequency in the base band spectrum, which is proportional to the radial distance of the target to the radar. According to the respective literature, e.g. in [Spa02], the time running during each ramp is labeled "fast time" and the time between the chirps is denoted as "slow time". For each of these "times", one FFT is performed, as can be seen schematically in the lower image of Fig. A.2, for the rows and columns. This 2D-FFT leads to a range-Doppler map, resp. range-velocity map, showing the targets as peaks in the spectrum. Before the FFTs are performed, a window function is used on the signals to compensate for the non-ideal periodic repetition of the signal and the discontinuities originating from it. Mainly, Kaiser and Hann window functions were used in this work. After the windowing had been applied, the FFTs with proper zero-padding factors are performed. These factors depend on the physical resolution of the radar, which is itself dependent on the modulation settings of the transmitted ramps. Usually, a factor of at least two is feasible. After the 2D spectrum is obtained, for each target, another FFT can be performed over the MIMO's virtual array (VA) antenna elements, in order to obtain a direction of arrival (DOA) spectrum. For this,

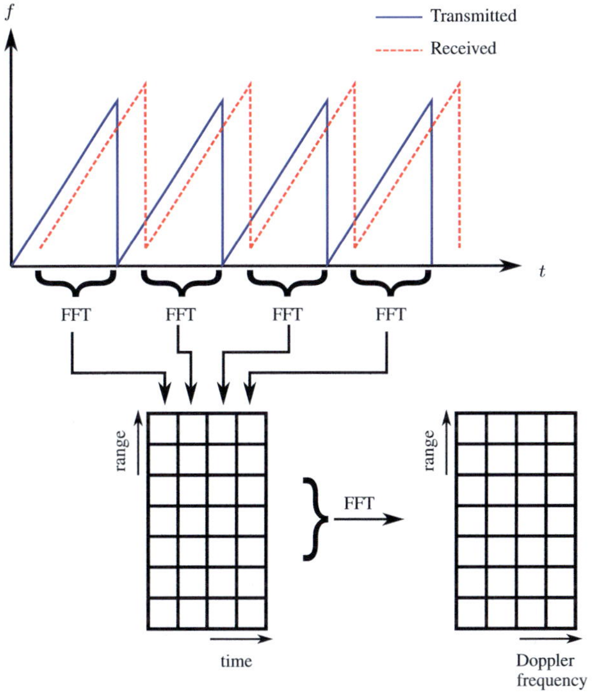

Figure A.2: Signal processing scheme of one frame comprising 4 chirps. Top: Schematic transmitted and received radio frequency (RF) FMCW chirps with range and Doppler shift over time. Bottom: 2D-fast Fourier transform (FFT) as a combination of two single FFTs. First, FFTs are performed over the samples of each ramp (colums). Then, a second FFT is performed over all ramps for each sample (rows). Source: [Ott12].

the complex values of the target's peak in the range-velocity map are utilized. This procedure is described in Section 2.3.2.

A.2 FMCW Modulation Parameters

In Chapters 4 to 6, the waveguide antenna from Fig. 3.2 was used in combination with the following modulation parameters for performing the measurements[1]:

- Number of Tx antennas: $N_{tx} = 8$
- Number of Rx antennas: $N_{rx} = 8$
- Center frequency: $f_c = 77\,\text{GHz}$
- Bandwidth: $B = 2\,\text{GHz}$
- Ramp time: $T_r = 100\,\mu\text{s}$
- Time between ramps: $T_{wr} = 18\,\mu\text{s}$
- Sampling rate: $S_r = 8\,\text{MHz}$
- Ramps per frame and Tx: $N_r = 50$
- Number of frames of one measurement per Tx: $N_m = 10$
- Time between the measurements: $T_{wm} = 100\,\mu\text{s}$

In Chapter 7, the redesigned waveguide antenna from Fig. 3.7 was used in combination with the following modulation parameters for performing the measurements[1]:

- Number of Tx antennas: $N_{tx} = 8$
- Number of Rx antennas: $N_{rx} = 8$
- Center frequency: $f_c = 77\,\text{GHz}$
- Bandwidth: $B = 1\,\text{GHz}$
- Ramp time: $T_r = 29\,\mu\text{s}$
- Time between ramps: $T_{wr} = 5\,\mu\text{s}$

[1] The bandwidth is given for the complete ramp time including the time between ramps, so the actually used bandwidth is slightly smaller

- Sampling rate: $S_r = 17.9\,\text{MHz}$
- Ramps per frame and Tx: $N_r = 64$
- Number of frames of one measurement per Tx: $N_m = 1$

Abbreviations and Acronyms

Adam	adaptive moment estimation
ADAS	advanced driver assistance systems
ADC	analog-to-digital converter
AOA	angle of arrival
AOI	angle of incidence
AWGN	additive white Gaussian noise
BSA	backscatter alignment
CAGR	compound annual growth rate
CCW	counterclockwise
CFAR	constant false alarm rate
CNN	convolutional neural network
co-pol	co-polarized
CST	CST MICROWAVE STUDIO®
DBF	digital beam forming
DGPS	differential global positioning system
DOA	direction of arrival
EM	electromagnetic
ESPRIT	estimation of signal parameters via rotational invariance technique
FDM	frequency-division multiplex
FFT	fast Fourier transform
FMCW	frequency-modulated continuous-wave

FOV	field of view
FPGA	field programmable gate array
GCS	geographic coordinate system
GCT	general calibration technique
GO	geometrical optics
GPR	ground penetrating radar
GPS	global positioning system
GT	ground truth
HH	horizontal-horizontal
HV	horizontal-vertical
I/O	input/output
IACT	isolated antenna calibration technique
kNN	k-nearest neighbor
LHC	left-hand circular
LHCP	left hand circular polarized
LNA	low noise amplifier
LOS	line-of-sight
MIMO	multiple-input, multiple-output
MLP	multilayer perceptron
MMIC	monolithic microwave integrated circuit
MMW	millimeter wave
MNIST	modified national institute of standards and technology
MUSIC	multiple signal classification
OFDM	orthogonal frequency-division multiplexing
PA	power amplifier
PC	personal computer

PEC	perfect electrically conducting
R-CNN	region-based convolutional neural network
RCS	radar cross section
ReLU	rectified linear unit
RF	radio frequency
RGB	red-green-blue
RHC	right-hand circular
RHCP	right hand circular polarized
ROI	region of interest
Rx	receiver
SAR	synthetic aperture radar
SD	standard deviation
SGD	stochastic gradient descent
SLA	stereolithography apparatus
SLS	side lobe suppression
S-matrix	scattering matrix
SNR	signal-to-noise ratio
SRGCT	single-reference general calibration technique
TDM	time-division multiplexing
TEM	transverse electromagnetic
TOF	time of flight
t-SNE	t-distributed stochastic neighbor embedding
Tx	transmitter
ULA	uniform linear array
VA	virtual array
VH	vertical-horizontal
VV	vertical-vertical

WGS 84	world geodetic system 1984
x-pol	cross-polarized
ZPF	zero-padding factor

Bibliography

[Alb+04] V. Alberga et al. "Potential of Coherent Decompositions in SAR Polarimetry and Interferometry". In: *IEEE International Geoscience and Remote Sensing Symposium (IGARSS)*. Vol. 3. Sept. 2004, 1792–1795 vol.3. DOI: 10.1109/IGARSS.2004.1370682.

[Alb04] V. Alberga. "Comparison of Polarimetric Methods in Image Classification and SAR Interferometry Applications". PhD thesis. Microwaves and Radar Institute at DLR, 2004.

[And+13] M. Andres et al. "3D Detection of Automobile Scattering Centers Using UWB Radar Sensors at 24/77 GHz". In: *IEEE Aerospace and Electronic Systems Magazine* 28.3 (Mar. 2013), pp. 20–25. ISSN: 0885-8985. DOI: 10.1109/MAES.2013.6495649.

[AS13] J. Adametz and L. P. Schmidt. "Threat Object Classification with a Close Range Polarimetric Imaging System by Means of H-alpha Decomposition". In: *European Radar Conference (EuRAD)*. Oct. 2013, pp. 77–80.

[Avi13] Avionics Department. *Electronic Warfare and Radar Systems Engineering Handbook*. 4th ed. Naval Air Warfare Center Weapons Division, Oct. 2013.

[Bal05] C. A. Balanis. *Antenna Theory: Analysis and Design*. 3rd ed. Antenna Theory: Analysis and Design v. 1. John Wiley & Sons, 2005. ISBN: 9780471667827.

[Bar73] D. E. Barrick. *FM/CW Radar Signals and Digital Processing*. Tech. rep. National Oceanic and Atmospheric Administration (NOAA), July 1973.

[BC01] V. N. Bringi and V. Chandrasekar. *Polarimetric Doppler Weather Radar: Principles and Applications*. Cambridge University Press, 2001. ISBN: 9780521623841.

[BC15] D. Belgiovane and C. C. Chen. "Radar Scattering Properties of Bicycles at 77 GHz". In: *IEEE International Symposium on Antennas and Propagation USNC/URSI National Radio Science Meeting*. July 2015, pp. 502–503. DOI: 10.1109/APS.2015.7304637.

[BC16] D. Belgiovane and C. C. Chen. "Bicycles and Human Riders Backscattering at 77 GHz for Automotive Radar". In: *10th European Conference on Antennas and Propagation (EuCAP)*. Apr. 2016, pp. 1–5. DOI: 10.1109/EuCAP.2016.7481649.

[BFF06] D. Bliss, K. Forsythe, and G. Fawcett. "MIMO Radar: Resolution, Performance, and Waveforms". In: *The Fourteenth Annual Workshop on Adaptive Sensor Array Processing, MIT Lincoln Laboratory*. June 2006.

[BH10] M. E. Bouzouraa and U. Hofmann. "Fusion of Occupancy Grid Mapping and Model Based Object Tracking for Driver Assistance Systems Using Laser and Radar Sensors". In: *IEEE Intelligent Vehicles Symposium*. June 2010, pp. 294–300. DOI: 10.1109/IVS.2010.5548106.

[Bha+98] R. Bhalla et al. "3D Scattering Center Representation of Complex Targets Using the Shooting and Bouncing Ray Technique: A Review". In: *IEEE Antennas and Propagation Magazine* 40.5 (Oct. 1998), pp. 30–39. ISSN: 1045-9243. DOI: 10.1109/74.735963.

[Bli+09] D. W. Bliss et al. "GMTI MIMO Radar". In: *Waveform Diversity and Design Conference*. Ed. by Institute of Electrical and Electronics Engineers. 2009.

[Bli08] D. W. Bliss. *Multiple-Antenna Techniques for Wireless Communications*. Tech. rep. MIT Lincoln Laboratory, 2008.

[Boe07] W. M. Boerner. *BASICS OF SAR POLARIMETRY I & II*. Report. UIC-ECE Communications, Sensing & Navigation Laboratory 900 W, 2007.

[BW99] M. Born and E. Wolf. *Principles of Optics: Electromagnetic Theory of Propagation, Interference and Diffraction of Light*. 7th ed. Cambridge University Press, 1999. ISBN: 0521642221.

[Che+16] S. Chen et al. "Unsupervised Classification Based on the Logarithmic Circular Polarization Ratio Parameter for Hybrid Polarimetric SAR". In: *IEEE International Geoscience and Remote Sensing Symposium (IGARSS)*. July 2016, pp. 979–982. DOI: 10.1109/IGARSS.2016.7729248.

[Che17] K. Cheromcha. "Self-Driving Cars Are Tripping Balls On Lidar, Basically". In: *The Drive* (May 26, 2017).

[Clo09] S. R. Cloude. *Polarisation: Applications in Remote Sensing*. Oxford Univeristy Press, 2009. ISBN: 9780199569731.

[CP97] S. R. Cloude and E. Pottier. "An Entropy Based Classification Scheme for Land Applications of Polarimetric SAR". In: *IEEE Transactions on Geoscience and Remote Sensing* 35.1 (Jan. 1997), pp. 68–78. ISSN: 0196-2892. DOI: 10.1109/36.551935.

[Dic+16] J. Dickmann et al. "Automotive Radar the Key Technology for Autonomous Driving: From Detection and Ranging to Environmental Understanding". In: *IEEE Radar Conference (RadarConf)*. May 2016, pp. 1–6. DOI: 10.1109/RADAR.2016.7485214.

[Die+11] F. Diewald et al. "Radar-Interference-Based Bridge Identification for Collision Avoidance Systems". In: *IEEE Intelligent Vehicles Symposium (IV)*. June 2011, pp. 113–118. DOI: 10.1109/IVS.2011.5940422.

[EA83] G. A. Emmons and P. M. Alexander. *Polarization Scattering Matrices for Polarimetric Radar*. Tech. rep. Advanced Sensors Directorate US Army Missile Laboratory, Mar. 1983.

[Fen+15] X. Feng et al. "Combination of H-Alpha Decomposition and Migration for Enhancing Subsurface Target Classification of GPR". In: *IEEE Transactions on Geoscience and Remote Sensing* 53.9 (Sept. 2015), pp. 4852–4861. ISSN: 0196-2892. DOI: 10.1109/TGRS.2015.2411572.

[FLS66] R. P. Feynman, R. B. Leighton, and M. Sands. *The Feynman Lectures on Physics, Vol. III, Quantum Mechanics*. Addison-Wesley, 1966.

[Fuk90] K. Fukunaga. *Introduction to Statistical Pattern Recognition*. 2nd ed. Academic Press, 1990.

[GBC16] I. Goodfellow, Y. Bengio, and A. Courville. *Deep Learning*. http://www.deeplearningbook.org. MIT Press, 2016.

[GGG55] W. Gronwald, W. Grossmann, and T. Gerardy. *C. F. Gauss und die Landesvermessung in Niedersachsen*. Niedersächsisches Landesvermessungsamt, Hannover, 1955.

[Gir+16] R. Girshick et al. "Region-Based Convolutional Networks for Accurate Object Detection and Segmentation". In: *IEEE Transactions on Pattern Analysis and Machine Intelligence* 38.1 (Jan. 2016), pp. 142–158. ISSN: 0162-8828. DOI: 10.1109/TPAMI.2015.2437384.

[Gol17] D. H. Goldstein. *Polarized Light, Third Edition*. CRC Press, 2017. ISBN: 9781439830413.

[Gol98] P. F. Goldsmith. *Quasioptical Systems: Gaussian Beam Quasioptical Propogation and Applications*. IEEE Press Series on RF and Microwave Technology. Wiley, 1998. ISBN: 9780780334397.

[Gra17] Grand View Research. *Automotive Radar Market*. Market Estimates & Trend Analysis. Grand View Research Inc., USA, Aug. 2017.

[Haj01] I. Hajnsek. "Inversion of Surface Parameters Using Polarimetric SAR". PhD thesis. Friedrich-Schiller-Universität Jena, 2001.

[HE07] J. Huang and J. A. Encinar. *Reflectarray Antennas*. IEEE Press Series on Electromagnetic Wave Theory. Wiley, 2007. ISBN: 9780470178768.

[HPC01] I. Hajnsek, K. P. Papathanassiou, and S. R. Cloude. "Removal of Additive Noise in Polarimetric Eigenvalue Processing". In: *International Geoscience and Remote Sensing Symposium (IGARSS)*. Vol. 6. 2001, pp. 2778–2780. DOI: 10.1109/IGARSS.2001.978160.

[HPC02] I. Hajnsek, K. P. Papathanassiou, and S. R. Cloude. "Surface Parameter Estimation Using Fully Polarimetric L- and P-band Radar Data". In: *Retrieval of Bio- and Geo-Physical Parameters from SAR Data for Land Applications*. Ed. by A. Wilson and S. Quegan. Vol. 475. ESA Special Publication. Jan. 2002, pp. 159–164.

[Hua+11] Y. Huang et al. "FMCW Based MIMO Imaging Radar for Maritime Navigation". In: *Progress In Electromagnetics Research*. Vol. 115. Apr. 2011, pp. 327–342.

[KB14] D. P. Kingma and J. Ba. "Adam: A Method for Stochastic Optimization". In: *ArXiv e-prints* (Dec. 2014). arXiv: `1412.6980 [cs.LG]`.

[Kno12] E. F. Knott. *Radar Cross Section Measurements*. Springer US, 2012. ISBN: 9781468499049.

[Kra02] H. J. Kramer. *Observation of the Earth and Its Environment - Survey of Missions and Sensors*. 4th ed. Springer, Berlin, 2002.

[Kra88] J. D. Kraus. *Antennas*. Electrical Engineering Series. McGraw–Hill, 1988. ISBN: 9780070354227.

[LaP17a] M. LaPedus. "Here Comes High-Res Car Radar". In: *Semiconductor Engineering* (Nov. 20, 2017).

[LaP17b] M. LaPedus. "Radar Versus LiDAR". In: *Semiconductor Engineering* (Oct. 23, 2017).

[LB90] B. J. Lipa and D. E. Barrick. *FMCW Signal Processing*. Mirage Systems, 1990.

[Lec+98] Y. Lecun et al. "Gradient-Based Learning Applied to Document Recognition". In: *Proceedings of the IEEE* 86.11 (Nov. 1998), pp. 2278–2324. ISSN: 0018-9219. DOI: `10.1109/5.726791`.

[LP09] J. S. Lee and E. Pottier. *Polarimetric Radar Imaging: From Basics to Applications*. Optical Science and Engineering. CRC Press, 2009. ISBN: 9781420054989.

[LS08] J. Li and P. Stoica. *MIMO Radar Signal Processing*. John Wiley & Sons, 2008. ISBN: 9780470391433.

[Mar12] A. Marino. "A New Target Detector Based on Geometrical Perturbation Filters for Polarimetric Synthetic Aperture Radar (POLSAR)". PhD thesis. 2012.

[MH08] L. van der Maaten and G. Hinton. "Visualizing Data using t-SNE". In: *Journal of Machine Learning Research* 9 (2008). Ed. by Y. Bengio.

[Mot06] H. Mott. *Remote Sensing with Polarimetric Radar*. John Wiley & Sons, 2006. ISBN: 9780470079805.

[NH10] V. Nair and G. E. Hinton. "Rectified Linear Units Improve Restricted Boltzmann Machines". In: *Proceedings of the 27th International Conference on Machine Learning (ICML-10)*. Omnipress, 2010, pp. 807–814.

[Ott12] T. Otto. *Principle of FMCW Radars*. Ed. by TU Delft (ATMOS). July 28, 2012.

[Ram+14] N. Ramli et al. "Design of an Aperture-Coupled Frequency-Reconfigurable Microstrip Stacked Array Antenna for LTE and WiMAX Applications". In: *ISRN Communications and Networking* (2014). Ed. by ISRN Communications and Networking. DOI: 10.1155/2014/154518.

[Ran07] R. K. Raney. "Hybrid-Polarity SAR Architecture". In: *IEEE Transactions on Geoscience and Remote Sensing* 45.11 (Nov. 2007), pp. 3397–3404. ISSN: 0196-2892. DOI: 10.1109/TGRS.2007.895883.

[Roo+18] F. Roos et al. "Enhancement of Doppler Unambiguity for Chirp-Sequence Modulated TDM-MIMO Radars". In: *IEEE MTT-S International Conference on Microwaves for Intelligent Mobility (ICMIM)*. Apr. 2018.

[Rus17] D. Rusev. "Using Convolutional Neural Networks for Object Recognition in Polarimetric Radar Images of Traffic Scenarios". MA thesis. Karlsruhe Institute of Technology, 2017.

[RV17] G. Rudolph and U. Voelzke. "Three Sensor Types Drive Autonomous Vehicles". In: *Sensors Magazine* (Nov. 10, 2017).

[Ryt+15] R. Rytel-Andrianik et al. "Simple X-Band Polarimetric Micro-Doppler Analyses of Ground Moving Targets". In: *Signal Processing Symposium (SPSympo)*. June 2015, pp. 1–4. DOI: 10.1109/SPS.2015.7168301.

[Sag17] A. Sagainov. "Machine Learning Algorithms for Classification with Polarimetric Radar". MA thesis. Karlsruhe Institute of Technology, 2017.

[SBW08] K. Schuler, D. Becker, and W. Wiesbeck. “Extraction of Virtual Scattering Centers of Vehicles by Ray-Tracing Simulations”. In: *IEEE Transactions on Antennas and Propagation* 56.11 (Nov. 2008), pp. 3543–3551. ISSN: 0018-926X. DOI: 10.1109/TAP.2008.2005436.

[Sch+12] C. M. Schmid et al. “Motion Compensation and Efficient Array Design for TDMA FMCW MIMO Radar Systems”. In: *6th European Conference on Antennas and Propagation (EUCAP).* Mar. 2012, pp. 1746–1750. DOI: 10.1109/EuCAP.2012.6206605.

[Sch98] R. Schneider. “Modellierung der Wellenausbreitung für ein bildgebendes Kfz-Radar”. PhD thesis. Universität Fridericiana Karlsruhe, May 1998.

[SD91] L. L. Scharf and C. Demeure. *Statistical Signal Processing: Detection, Estimation, and Time Series Analysis.* Addison-Wesley series in electrical and computer engineering. Addison-Wesley Publishing Company, 1991. ISBN: 9780201190380.

[SDW00] R. Schneider, D. Didascalou, and W. Wiesbeck. “Impact of Road Surfaces on Millimeter-Wave Propagation”. In: *IEEE Transactions on Vehicular Technology* 49.4 (July 2000), pp. 1314–1320. ISSN: 0018-9545. DOI: 10.1109/25.875249.

[Sha17] G. Shar. “Self-Driving Market Worth $83 Billion by 2025 with AI, Cloud & Mobility”. In: *AUTO Connected Car News* (May 9, 2017).

[Shi98] E. Shih-Syou Li. “Millimeter-Wave Polarimetric Radar System as an Advanced Vehicle Control and Warning Sensor”. PhD thesis. The University of Michigan, Oct. 1998.

[Sko02] M. Skolnik. *Introduction to Radar Systems.* 3rd ed. McGraw-Hill Education, 2002. ISBN: 9780072881387.

[SLD17] E. Shelhamer, J. Long, and T. Darrell. “Fully Convolutional Networks for Semantic Segmentation”. In: *IEEE Transactions on Pattern Analysis and Machine Intelligence* 39.4 (Apr. 2017), pp. 640–651. ISSN: 0162-8828. DOI: 10.1109/TPAMI.2016.2572683.

[Spa02] T. Sparr. "Time-Frequency Analysis for Radar Signal Processing". In: *5th Nordic Signal Processing Symposium (NORSIG)* (2002).

[Ste00] S. Stergiopoulos. *Advanced Signal Processing Handbook: Theory and Implementation for Radar, Sonar, and Medical Imaging Real Time Systems*. Electrical Engineering & Applied Signal Processing Series. CRC Press, 2000. ISBN: 9781420037395.

[Str05] G. Strang. *Linear Algebra and Its Applications*. 4th ed. Brooks, 2005. ISBN: 0030105676.

[SUT90] K. Sarabandi, F. T. Ulaby, and M. A. Tassoudji. "Calibration of Polarimetric Radar Systems With Good Polarization Isolation". In: *IEEE Transactions on Geoscience and Remote Sensing* 28.1 (Jan. 1990), pp. 70–75. ISSN: 0196-2892. DOI: 10.1109/36.45747.

[Tre18] L. Trego. "ADAS market worth $67.43 billion by 2025". In: *Autonomous Vehicle Technology Magazine* (Feb. 14, 2018).

[TRF15] Z. Tong, R. Renter, and M. Fujimoto. "Fast Chirp FMCW Radar in Automotive Applications". In: *IET International Radar Conference*. Oct. 2015, pp. 1–4. DOI: 10.1049/cp.2015.1362.

[Ulr08] P. Ulrich. "Charakterisierung des Radarstreuverhaltens von Objekten im automobilen Umfeld bei 77 GHz". Diploma thesis. Julius-Maximilians-Universität Würzburg, 2008.

[Wen+07] P. Wenig et al. "System Design of a 77 GHz Automotive Radar Sensor with Superresolution DOA Estimation". In: *International Symposium on Signals, Systems and Electronics*. July 2007, pp. 537–540. DOI: 10.1109/ISSSE.2007.4294531.

[Whi+91] M. W. Whitt et al. "A General Polarimetric Radar Calibration Technique". In: *IEEE Transactions on Antennas and Propagation* 39.1 (Jan. 1991), pp. 62–67. ISSN: 0018-926X. DOI: 10.1109/8.64436.

[WK91] W. Wiesbeck and D. Kähny. "Single Reference, Three Target Calibration and Error Correction for Monostatic, Polarimetric Free Space Measurements". In: *Proceedings of the IEEE* 79.10 (Oct. 1991), pp. 1551–1558. ISSN: 0018-9219. DOI: 10.1109/5.104229.

[WWS08] P. Wenig, R. Weigel, and M. Schneider. "A Dielectric Lens Antenna for Digital Beamforming and Superresolution DOA Estimation in 77 GHz Automotive Radar". In: *International ITG Workshop on Smart Antennas*. Feb. 2008, pp. 184–189. DOI: 10.1109/WSA.2008.4475557.

List of Own Publications

[1] T. Visentin, W. Keusgen, and R. Weiler. "Dual-Polarized Square-Shaped Offset-Fed Reflectarray Antenna with High Gain and High Bandwidth in the 60 GHz Domain". In: *9th European Conference on Antennas and Propagation (EuCAP)*. May 2015, pp. 1–5.

[2] T. Visentin, J. Hasch, and T. Zwick. "Analysis of Multipath and DOA Detection Using a Fully Polarimetric Automotive Radar". In: *20th European Microwave Week (EuMW)*. Oct. 2017.

[3] T. Visentin, J. Hasch, and T. Zwick. "Calibration of a Fully Polarimetric 8x8 MIMO FMCW Radar System at 77 GHz". In: *11th European Conference on Antennas and Propagation (EuCAP)*. Mar. 2017.

[4] T. Visentin, J. Hasch, and T. Zwick. "Polarimetric RCS Measurements of Selected Two-Wheeled Vehicles for Automotive Radar". In: *20th European Microwave Week (EuMW)*. Oct. 2017.

[5] T. Visentin, W. Keusgen, and R. Weiler. "Reflector Having an Electronic Circuit and Antenna Device Having a Reflector". 2017. Pending patent application.

[6] T. Visentin et al. "Classification of Objects in Polarimetric Radar Images Using CNNs at 77 GHz". In: *IEEE Asia Pacific Microwave Conference (APMC)*. Nov. 2017, pp. 356–359.

[7] T. Visentin, J. Hasch, and T. Zwick. "Analysis of Multipath and DOA Detection Using a Fully Polarimetric Automotive Radar". In: *International Journal of Microwave and Wireless Technologies (IJMWT)* (Apr. 3, 2018). Ed. by Cambridge University Press.

[8] T. Visentin et al. "Cross-Polarized Planar Reflector for Polarimetric Radar Calibration at 77 GHz". In: *11th German Microwave Conference (GeMIC)*. Mar. 2018.

Forschungsberichte aus dem Institut für Höchstfrequenztechnik und Elektronik (IHE) der Universität Karlsruhe (TH)
(ISSN 0942-2935)

Herausgeber: Prof. Dr.-Ing. Dr. h.c. Dr.-Ing. E.h. mult. Werner Wiesbeck

Die Bände 1 (1992) bis 55 (2008) der Schriftenreihe können über das Institut Hochfrequenztechnik und Elektronik bestellt werden (https://www.ihe.kit.edu).

Fortführung als:

Karlsruher Forschungsberichte aus dem Institut für Hochfrequenztechnik und Elektronik
(ISSN 1868-4696)

Herausgeber: Prof. Dr.-Ing. Thomas Zwick

Band 55 Sandra Knörzer
Funkkanalmodellierung für OFDM-Kommunikationssysteme bei Hochgeschwindigkeitszügen (2009)
ISBN 978-3-86644-361-7

Band 56 Thomas Fügen
Richtungsaufgelöste Kanalmodellierung und Systemstudien für Mehrantennensysteme in urbanen Gebieten (2009)
ISBN 978-3-86644-420-1

Band 57 Elena Pancera
Strategies for Time Domain Characterization of UWB Components and Systems (2009)
ISBN 978-3-86644-417-1

Band 58 Jens Timmermann
Systemanalyse und Optimierung der Ultrabreitband-Übertragung (2010)
ISBN 978-3-86644-460-7

Band 59 Juan Pontes
Analysis and Design of Multiple Element Antennas for Urban Communication (2010)
ISBN 978-3-86644-513-0

Band 60 Andreas Lambrecht
True-Time-Delay Beamforming für ultrabreitbandige Systeme hoher Leistung (2010)
ISBN 978-3-86644-522-2

Band 61 Grzegorz Adamiuk
Methoden zur Realisierung von dual-orthogonal, linear polarisierten Antennen für die UWB-Technik (2010)
ISBN 978-3-86644-573-4

Band 62 Jutta Kühn
AlGaN/GaN-HEMT Power Amplifiers with Optimized Power-Added Efficiency for X-Band Applications (2011)
ISBN 978-3-86644-615-1

Band 63 Małgorzata Janson
Hybride Funkkanalmodellierung für ultrabreitbandige MIMO-Systeme (2011)
ISBN 978-3-86644-639-7

Band 64 Mario Pauli
Dekontaminierung verseuchter Böden durch Mikrowellenheizung (2011)
ISBN 978-3-86644-696-0

Band 65 Thorsten Kayser
Feldtheoretische Modellierung der Materialprozessierung mit Mikrowellen im Durchlaufbetrieb (2011)
ISBN 978-3-86644-719-6

Band 66 Christian Andreas Sturm
Gemeinsame Realisierung von Radar-Sensorik und Funkkommunikation mit OFDM-Signalen (2012)
ISBN 978-3-86644-879-7

Band 67 Huaming Wu
Motion Compensation for Near-Range Synthetic Aperture Radar Applications (2012)
ISBN 978-3-86644-906-0

Band 68 Friederike Brendel
Millimeter-Wave Radio-over-Fiber Links based on Mode-Locked Laser Diodes (2013)
ISBN 978-3-86644-986-2

Band 69 Lars Reichardt
Methodik für den Entwurf von kapazitätsoptimierten Mehrantennensystemen am Fahrzeug (2013)
ISBN 978-3-7315-0047-6

Band 70 Stefan Beer
Methoden und Techniken zur Integration von 122 GHz Antennen in miniaturisierte Radarsensoren (2013)
ISBN 978-3-7315-0051-3

Band 71 Łukasz Zwirełło
Realization Limits of Impulse-Radio UWB Indoor Localization Systems (2013)
ISBN 978-3-7315-0114-5

Band 72 Xuyang Li
Body Matched Antennas for Microwave Medical Applications (2014)
ISBN 978-3-7315-0147-3

Band 73 Sebastian Diebold
Transistor- und Leitungsmodellierung zum Entwurf von monolithisch integrierten Leistungsverstärkern für den hohen Millimeterwellen-Frequenzbereich (2014)
ISBN 978-3-7315-0161-9

Band 74 Christian Rusch
Integrierte, planare Leckwellenantennen für 3D-Millimeterwellen-Radarsysteme basierend auf dem holografischen Prinzip (2014)
ISBN 978-3-7315-0234-0

Band 75 Marlene Harter
Dreidimensional bildgebendes Radarsystem mit digitaler Strahlformung für industrielle Anwendungen (2014)
ISBN 978-3-7315-0249-4

Band 76 Michael A. Baldauf
Abhängigkeit der Exposition von der Zellgröße beim Mobilfunk unter Gewährleistung der Versorgung (2015)
ISBN 978-3-7315-0308-8

Band 77 Alicja Ossowska
Highly Resolved Synthetic Aperture Radar with Beam Steering (2015)
ISBN 978-3-7315-0315-6

Band 78 Małgorzata Dominika Brzeska
RF Modelling and Characterization of Tyre Pressure Sensors and Vehicle Access Systems (2015)
ISBN 978-3-7315-0348-4

Band 79 Ulrich Lewark
Aktive Frequenzvervielfacher zur Signalerzeugung im Millimeter- und Submillimeterwellen Frequenzbereich (2015)
ISBN 978-3-7315-0354-5

Band 80 Kai-Philipp Walter Pahl
Distributed Transformers for Broadband Monolithic Millimeter-Wave Integrated Power Amplifiers (2015)
ISBN 978-3-7315-0409-2

Band 81 Serdal Ayhan
Hochgenaue radarbasierte Abstandsmessung mit geführter Wellenausbreitung (2016)
ISBN 978-3-7315-0433-7

Band 82 Yoke Leen Sit
MIMO OFDM Radar-Communication System with Mutual Interference Cancellation (2017)
ISBN 978-3-7315-0599-0

Band 83 Steffen Scherr
FMCW-Radarsignalverarbeitung zur Entfernungsmessung mit hoher Genauigkeit (2017)
ISBN 978-3-7315-0607-2

Band 84 Tom Schipper
Modellbasierte Analyse des Interferenzverhaltens von Kfz-Radaren (2017)
ISBN 978-3-7315-0639-3

Band 85 Malyhe Jalilvand
Application-Specific Broadband Antennas for Microwave Medical Imaging (2017)
ISBN 978-3-7315-0664-5

Band 86 Benjamin Göttel
Millimeterwellen On-Chip Antennensysteme für die Integration in SoC Applikationen (2017)
ISBN 978-3-7315-0667-6

Band 87 Christian Arnold
Im Orbit einstellbare Ausgangsfilter und -multiplexer (2017)
ISBN 978-3-7315-0722-2

Band 88 Tobias Mahler
Synthese kapazitätsoptimierter Antennensysteme mit messtechnischer Verifikation (2018)
ISBN 978-3-7315-0737-6

Band 89 Daniel Müller
RF Probe-Induced On-Wafer Measurement Errors in the Millimeter-Wave Frequency Range (2018)
ISBN 978-3-7315-0822-9

Band 90 Tristan Visentin
Polarimetric Radar for Automotive Applications (2019)
ISBN 978-3-7315-0888-5